■ Two Representations

Then and Now

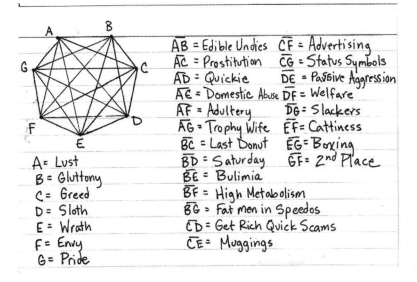

A = Lust
B = Gluttony
C = Greed
D = Sloth
E = Wrath
F = Envy
G = Pride

$\overline{AB}$ = Edible Undies
$\overline{AC}$ = Prostitution
$\overline{AD}$ = Quickie
$\overline{AE}$ = Domestic Abuse
$\overline{AF}$ = Adultery
$\overline{AG}$ = Trophy Wife
$\overline{BC}$ = Last Donut
$\overline{BD}$ = Saturday
$\overline{BE}$ = Bulimia
$\overline{BF}$ = High Metabolism
$\overline{BG}$ = Fat men in Speedos
$\overline{CD}$ = Get Rich Quick Scams
$\overline{CE}$ = Muggings
$\overline{CF}$ = Advertising
$\overline{CG}$ = Status Symbols
$\overline{DE}$ = Passive Aggression
$\overline{DF}$ = Welfare
$\overline{DG}$ = Slackers
$\overline{EF}$ = Cattiness
$\overline{EG}$ = Boxing
$\overline{GF}$ = 2nd Place

"We're all going to Hell."

Posted by Jessica Hagy, January 29, 2007. Available at: http://indexed.blogspot.com/2007/01/were-all-going-to-hell.html.

1

STANDARDS
AND
THEIR
STORIES

STANDARDS AND THEIR STORIES

HOW QUANTIFYING, CLASSIFYING, AND FORMALIZING PRACTICES SHAPE EVERYDAY LIFE

Edited by **Martha Lampland**
and **Susan Leigh Star**

Cornell University Press

Ithaca & London

First published 2009 by Cornell University Press
First printing, Cornell Paperbacks, 2009

Printed in the United States of America

Library of Congress Cataloging-in-Publication Data
Standards and their stories : how quantifying, classifying, and formalizing practices shape everyday
life / edited by Martha Lampland and Susan Leigh Star.
    p. cm.
  Includes bibliographical references and index.
  ISBN 978-0-8014-4717-4 (cloth : alk. paper) — ISBN 978-0-8014-7461-3 (pbk. : alk. paper)
  1. Standardization—Social aspects.   2. Standards—Social aspects.   3. Measurement—Social
aspects.   4. Classification—Social aspects.   I. Lampland, Martha, 1952–   II. Star, Susan Leigh,
1954–
  HD62.S785 2009
  389'.6—dc22   2008029675

Cornell University Press strives to use environmentally responsible suppliers
and materials to the fullest extent possible in the publishing of its books. Such
materials include vegetable-based, low-VOC inks and acid-free papers that are
recycled, totally chlorine-free, or partly composed of nonwood fibers. For further
information, visit our website at www.cornellpress.cornell.edu.

Cloth printing   10 9 8 7 6 5 4 3 2 1

Paperback printing   10 9 8 7 6 5 4 3 2

For those who labor long and hard to craft good and just standards, as well as for those who have suffered from their absence. On the one hand, the fight against the tyranny of structurelessness. On the other, the fallacy of one size fits all.

A WORD ON STATISTICS

Wislawa Szymborska

translated from the Polish by Joanna Trzeciak

Out of every hundred people,
those who always know better:
fifty-two.

Unsure of every step:
almost all the rest.

Ready to help,
if it doesn't take long:
forty-nine.

Always good,
because they cannot be otherwise:
four—well, maybe five.

Able to admire without envy:
eighteen.

Led to error
by youth (which passes):
sixty, plus or minus.

Those not to be messed with:
four-and-forty.

Living in constant fear
of someone or something:
seventy-seven.

Capable of happiness:
twenty-some-odd at most.

Harmless alone,
turning savage in crowds:
more than half, for sure.

Cruel
when forced by circumstances:
it's better not to know,
not even approximately.

Wise in hindsight:
not many more
than wise in foresight.

Getting nothing out of life except things:
thirty (though I would like to be wrong).

Balled up in pain
and without a flashlight in the dark:
eighty-three, sooner or later.

Those who are just:
quite a few, thirty-five.

But if it takes effort to understand:
three.

Worthy of empathy:
ninety-nine.

Mortal:
one hundred out of one hundred—
a figure that has never varied yet.

CONTENTS

7. ASCII Imperialism *177*

Daniel Pargman and Jacob Palme

ACKNOWLEDGMENTS

■ Two groups of people helped us create this book. The first was a faculty research study seminar at the University of California Humanities Research Institute, convened by the editors of this volume. We spent a term together considering "Historical and Interpretive Approaches to Standardization, Quantification and Formal Representation." The staff at the institute and David Theo Goldberg, its director, were wonderfully helpful. The attendees were Rogers Hall, Judith Treas, Geoffrey Bowker, Martin Lengweiler, L. K. (Mimi) Saunders, Theodore Porter, Janice Neri, Jean Lave, and ourselves. Important guests of the seminar were Kathleen Casey, Helen Verran, and Ole Dreier. The second group of people were our students in a co-taught seminar, "Standardization and Quantification," in the Science Studies Program at the University of California, San Diego: Ray-Shyng Chou, Sophia Efstathiou, Michael Evans, Carmel Finley, Laura Harkewicz, Lyn Headley, Brian Lindseth, Steven Luis, Eric Martin, Evelyn Range, Jesse Richmond, Tom Waidzunas, Kevin Walsh, and Alper Yalcinkaya. The work and leadership of Lawrence Busch, director of the Institute for Food and Agricultural Standards, Michigan State University, has provided continuing inspiration. Finally, we owe our editor, Peter Wissoker, a particular debt of thanks for understanding how the narratives surrounding standards are crucially important at this historical moment.

STANDARDS
AND
THEIR
STORIES

The seven deadly sins led by the devil. Animal symbols constituted a visual classificatory scheme depicting the seven deadly sins, visible on the flags, for example, a pig for lust, the head of a bear for gluttony, a hedgehog for greed, an ass for sloth, a wolf for wrath, and a lion for pride.

Anonymous artists, woodcut, published September 25, 1490.

RECKONING WITH STANDARDS

Susan Leigh Star and Martha Lampland

■ A friend, who has recently moved to the Netherlands, makes an appointment to see a U.S. tax preparer. She has no phone. She walks in to the tax office and schedules a time for the next day.

"What is your phone number, please?" asks the polite young man managing the office calendar.

"I don't have one."

"I'm sorry, I can't put your appointment in to the calendar without a phone number."

"Yes, but I don't have one."

Silence.

"Would you like me to make one up?" asks our friend.

"Oh, yes," sighs the calendar-filler, "that would be great."

"1-2-3-4-5-6-7," says my friend.

"Perfect!" the young man says. "The computer accepted that just fine. See you tomorrow!"

This short anecdote introduces one of the main themes of this book: how contemporary people interact with standardized forms, technologies, and conventions built into infrastructure. This compilation is one of a few papers and books beginning to analyze the contemporary view of this question, one that includes the growing place of all sorts of standards, formal and informal, in our everyday lives (Bingen and Busch 2006; Brunsson and Jacobsson and Associates 2000; Busch, 2000). This growth is apparent at the most minute level and at the most macro level.

In the grocery store, labels referring to standards blossom. At times even each *piece* of fruit is labeled with a number, referring to a particular farm or crop; often there is other information pointing to rules, standardized practices, or other techno-socio-agricultural constraints. Standards for the labeling of food are constantly changing, most recently including properties of its manufacture such as whether the food has been through a mill where nuts, wheat, or other allergens have also been processed. Sometimes in the United States, warnings of genetically modified organism (GMO) foodstuffs is to be found there; this is now standard in Europe. Standards for naming a product *organic* have recently been established in California; terms such as *natural* or *free range* are not standardized and, essentially, mean nothing. Outside of the market, the number and location of disabled parking spots are standardized and regulated by the Department of Motor Vehicles. This is a quick, light observation. A detailed study of even one market would reveals thousands of interlocking standards (and even more if it is one of a franchise of markets). How can we even approach this thicket?

This book considers a specific question: How have people dealt, in ordinary ways, with these millions of interlocking standards? Although the anecdote at the beginning of this chapter is meant to be amusing, it is, at the same time, deadly serious. These sorts of workarounds and stalling off computerized consequences are ubiquitous—work must get done, even though one size never fits all. The data that are missing when this happens are part of a vast domain of shadow work (Illich 1981) that can never be recovered. At the same time, these practices can be crucial to our understanding of how things really occur in the workplace. We hope to contribute here, in a modest way, to the dulling of the impulse to standardize everything that seems to grip modern organizations. We are not, in any sense, against standardizing—only against society's romance with it.

Our purpose here is hardly to compile a comprehensive history of standardization. Our goal is to show how standards are phenomena worthy of study in their own right, from multiple social scientific points of view. We hope to invite other studies of both the mundane and the arcane, the unconscious use of standards and numbers, and their very conscious use in intellectual development and research itself. Investing in forms is a cultural historical project, as is the increasing marginalization or deletion of content and residual categories (Thévenot 1984). The chapters included in this book (and the eclectic collage of examples interspersed) constitute our attempts to grapple with the phenomena of standardization and quantification in several domains: biology, public news media, food preparation, work and labor units, insurance, education, and everyday activities such as shopping.

Analytic Commonalities

To understand this romance better and to think about the human use, creation, and disuse of standards, we have built on the chapters here to analyze their commonalities. One of the results found across these chapters is that standards, as with all similar forms of compression and representations of actions:

- Are nested inside one another.
- Are distributed unevenly across the sociocultural landscape.
- Are relative to communities of practice; that is, one person's well-fitting standard may be another's impossible nightmare.
- Are increasingly linked to and integrated with one another across many organizations, nations, and technical systems.
- Codify, embody, or prescribe ethics and values, often with great consequences for individuals (consider standardized testing in schools, for example).

Let us consider each of these dimensions in turn.

Nested

When we refer to standards as being nested, we are speaking of the ways that they fit inside one another, somewhat like a set of Russian dolls (maitruska). Returning to the seemingly simple example at the beginning of this chapter, we can pull on a thread anywhere in the example and see that its implications are recursive through many systems. There are apparently tiny standards, such as the form for filling in the telephone number in the tax preparer's electronic calendar. Most people who visit a tax-preparation company in the United States have a telephone and know its number; however, for a variety of reasons, not all do. Americans who live abroad but still pay taxes to the United States may temporarily not have a phone; a homeless person may not have a phone, but may pay taxes and even require assistance in doing so; newcomers to an area may not have a standard arrangement for receiving telephone calls; and so on. The standard of *having a phone* is linked with *making an appointment,* which is linked with an inflexible, standardized *computerized calendar.* Not to belabor the point, but there are also medium-sized standards lurking in the background in a much larger, encompassing "nest"; for example, the U.S. tax code is so complexly standardized that most middle-class people pay US$300–1,000 to have someone else navigate it for them every April 15. Quite large standards and practices—what percentage of a person's income goes toward state and federal taxes, and how little a person can get away with paying—nest the small interaction with the calendar. Many very rich people pay no taxes; they have enough money to purchase tax shelters and other workarounds to the standard percentages.

Laurent Thévenot, in "Rules and Implements: Investment in Forms" (1984), argues that we are increasingly forced to make and use these sorts of standards and their attendant forms (now usually computerized, but with a sizable amount of paper still remaining). The very stuff of bureaucratic action is just such an investment in form. Content, such as a telephone number, may vary from instance to instance, but in fact the shape of the form becomes the primary human-capital investment. The flexibility of such linkages, as well as of each form, is variable. Options such as "Other" or a box labeled "Other forms of contact" would actually free up the way the form is driving the interaction. At the same time, however, the nested structure of the forms remains and is essentially not disturbed by a small number of such workarounds or residual spaces. Martin Lengwiler's inspection of the substandard human (chap. 4 in this volume) works inside the conception of a standard human, an object and set of events that is constantly being molded. Martha Lampland's work science standards (chap. 5 in this volume) are nested within a hierarchy of social inequities and a commitment to certain moral principles—principles that keep being renegotiated as work itself changes. Likewise, the privileging of chronological age is nested within structures of administration forms and rights, such as the ability to vote, drink, fight in the military, and drive (Judith Treas, chap. 3 in this volume). The formal techniques described by Florence Millerand and Geoffrey Bowker (chap. 6 in this volume), such as quantification, are nested within other standards in order to summarize information; metadata are nested within a whole system of standards.

Distributed Unevenly

With respect both to impact and obligation, sociotechnical standards are distributed unevenly. So, for example, most students in most Western countries must take standardized examinations at various stages of their schooling. This is a thorny, politically charged question. The very rich and the very poor, however, often escape the obligatory taking of the tests, or they have different relationships to it. Very rich youth may be educated in a way that is exempt from standardized testing (elite boarding schools outside the tests' jurisdiction, private tutors replacing classroom teaching standardized to the test, and so forth). Very poor young people may run away from school altogether; be educated in an environment that is not equipped to educate them about testing, rendering the test results moot; start work as children; or never achieve standard literacy through schooling. As Martha Lampland points out, the meaning of a standard, such as a work-hour, varies according to political regime and class position. When and where an individual is born matters greatly, as we see when we reach into the past to look for an uneven distribution of standards. The very definition of age is culturally variable, both unevenly distributed across historical periods and relative to the needs of states, labor pools, and who really counts.

Being able to speak English during the past years of computerization conveys a great privilege with respect to standardization. The implementation of most programs has until recently (and even now, although things are better) relied on a system of encoding, ASCII, that disproportionately disadvantaged people whose alphabet used non-ASCII characters. The examples from Swedish given by Daniel Pargman and Jacob Palme (chap. 7 in this volume) make clear the subtle but real advantages conveyed in qualities such as searchability on the Web and how the use of non-ASCII characters affects that.

Finally, for those of us who stood shocked as the CNN narratives about September 11, 2001, unfolded, we saw a disproportionate use of a standard set of images and the hardening of a story, reaching into the ways that news is made and acceptable narratives are constructed.[1]

Relative to the User and Communities of Practice (Social Worlds)

Following on the last point about uneven distribution, standard forms are also relative in their impact, meaning, and reach into individual and organizational lives. Standards, and the actions surrounding them, do not occur acontextually. There is always a kind of economy and ecology of standards surrounding any individual instance. Thus, what is benignly standard for one person at one time may be a barrier, or even a life-threatening occurrence, for another. For example, the act of presenting a passport in a standard gesture, in a standard format, works for millions of people much of the time. But, of course, some people are stateless, some states' legitimacy is questioned by other states, and some people (e.g., infants and prisoners) may be necessarily linked to others in order to enact standard citizenship. Steven Epstein (chap. 2 in this volume) speaks of different standards for different bodies, and Lengwiler (chap. 4 in this volume) talks about standard versus substandard lives (according to an insurance company). Throughout the book, this relative sense of standards is very clear. Millerand and Bowker (chap. 6 in this volume) note that standards are always relative to the infrastructure within/upon/sometimes against which they are implemented. The need for and the politics of metastandards thus arise—although the problem does not stop here; it is recursive.

Integrated

As we sit down, perhaps in the morning with a cup of tea or coffee, and answer our e-mail, we may read a greeting from a friend, a new deadline from a boss, or an argument from a student about a recent grade. Regardless of the particular words or emotional tones of the e-mails, in reading any of them, we use (invoke might be a better word) thousands of standards. For e-mail to function properly, these standards must be integrated one with the other, beginning with the source of access to the Internet (the service provider), software for presenting messages from many sources and in many formats, and telephonic and other carrier standards, and continuing right down to the machine code in the terminal or computer on the desktop and out to the Internet with its complex, evolving sets of handshakes and protocols (see Abbate 1999). The nature of this integration is profound, global (not, however, universal), increasing, and evolving. Social science theorists face new challenges in understanding exactly how this integration forms and drives action. For instance, when parents use cell phones to locate their teenage daughters and sons, is this a new form of surveil-

---

1. The first official meeting of the standards research group was held on Tuesday, September 11, 2001.

lance? How do families then configure themselves around the contact provided here? The older forms of checking up on and managing teen behavior on the part of parents included having them "telephone in" and meet curfews or having them remain within eyesight or in a chaperoned place. Do the caller identification numbers now enabled through a cell phone change how offspring manage information about their whereabouts? When some form of tracking becomes integrated with cell phones (such as a global positioning system, GPS) and family cars, does the resulting emotional ecology change the meaning of trust? We are beginning to study and weave answers to these sorts of questions (for example, Millerand and Bowker, chap. 6 in this volume); at the same time, the situation is moving very quickly.

Embody Ethics and Values

To standardize an action, process, or thing means, at some level, to screen out unlimited diversity. At times, it may mean to screen out even limited diversity. For example, despite the fact that transsexual and intersexual individuals have been a highly publicized, well-known aspect of modern culture for at least twenty years, almost all forms seeking demographic data have one binary choice, "M/F" (or male/female). And despite the fact that forms of partnership range from a single male and single female conjoined in one marriage for life to polyamorous arrangements with multiple genders and numbers of partners, most demographic forms ask "Married" (answer: "yes/no") or the functional equivalent. The silencing of "Other" choices here is a moral choice as well as a technical and data-collecting one. Where on a form do the transsexuals "go"? In traditional population census data (although this is changing dramatically in many places), where do people of racially mixed heritage (that is, all of us, if we carry this further) "go"? Often, individuals are forced to choose to self-silence some aspect of their lineage (see Bowker and Star 1999, chap. 5). Epstein (chap. 2 in this volume) speaks movingly of the ironies of resistance and the politics of representation in medical testing. When a person chooses one side of his or her heritage, it is often to redress inequalities conferred by the lesser status; other aspects of him or her self remain invisible. This invisibility is only one form of moral inscribing that derives from standardizing forms and processes. Others involve making things visible in a positive manner—such as including environmental data with economic assessments and including emotional stressors or physical danger with wages, where the more stressful or dangerous the job, the higher the wages, and this becomes formulaically part of pay. Sometimes, as Treas (chap. 3 in this volume) discusses, age conveys age-related benefits or honor—but just as often, it conveys discrimination. The wide range of values in design, use, and propagation of standard systems is another opportunity for social science/technology analysis. In the following section, we consider some of the ways the shadow work continues to be propagated.

Standards: Some Considerations of Invisibility

This book grows out of a research group devoted to thinking through three related phenomena: standardization, quantification, and formal representation.[2] These are phenomena, like the investment in form, that pervade modern life. For us as social scientists, one of the interesting aspects is that they have largely escaped consistent attention as sociocultural projects in themselves. The work of creating them is often invisible or deleted in descriptions of their development. Standardizing clothing sizes, developing indices of economic growth, creating computer databases, identifying the appropriate population for clinical trials in medicine, mandating testing in schools—all these procedures entail processes of standardization and quantification (and usually formal representation) (see Lynch 1991). Yet the standards, numbers, and models tend to be black boxes in their own right. They may be presented as secondary or epiphenomenal to the procedures of which they are a part: marketing for mass consumption, economic development policies, transmitting information, testing medical innovations, and supporting children's educational progress.

Talking with a production engineer, research scientist, classroom teacher, medical professional, or factory manager, we discover that their lives are filled with tasks designed to create standards or comply with existing standards. Associations such as the American Standards Institute or the International Standards Organisation (ISO) are familiar creatures in the current technological and production landscapes. Increasingly, humanities and the fine arts are filled with standards as well. Why, then, have they so often escaped social analysis?[3] Perhaps because many social scientists (ourselves included) fall into the taken-for-granted ease of seeing numbers and models, or specifications, as somehow "outside social order." Perhaps some of the neglect is due to the Byzantine politics of qualitative versus quantitative approaches in the social sciences.

Clearly, standards are also complexly related to quantification, formal modeling, and data mining, reuse, and classification. Another book or more would be needed adequately to describe the role of each of these. Quantification is the most developed historically and sociologically (see, e.g., Porter 1995; MacKenzie 2001). Some attention has been paid to formal modeling and its consequences, notably in the work of geographers and in the work of philosophers of biology (Wimsatt 1998; Griesemer 1990; see also Morgan and Morrison 1999). As a shorthand, and

2. *Quantification* is the representation of some action, being, or model through numbers. *Formal representations* are those not tied to a particular situation or set of empirical data but, rather, are a synthesis of data and a presentation of rules for combining and acting. These are often conveyed in visual form, as graphs, tables, or formulae. They may also be conveyed in narrative form, such as conventional sayings or standard characterizations of phenomena.

3. The exception here is the economic analysis of networked standards; see, for example, Paul David (1985).

to better focus, we use the term *standardizing* throughout the book. At many junctures, however, numbers and other formal tools play a critical role alongside standards, and we have tried to be aware of these moments.

Standardizing has become a central feature of social and cultural life in modernity.[4] The purpose of standardizing—to streamline procedures or regulate behaviors, to demand specific results, or to prevent harm—is rarely queried because it has come to be understood as a valuable and necessary, even if cumbersome, process. Certainly debates do take place over the extent or degree of standardization and especially about how and whether to measure the outcome of standardization (e.g., the nature and cultural bias of the IQ test and the Scholastic Aptitude Tests, SATs). But the question of whether to standardize (or quantify) *at all* is often suppressed. At times, it seems that standardizing overwhelms the primary activity— that is, the investment in form outweighs the performative content of the forms. Teachers, nurses, and psychotherapists, among others, criticize the increasing amount of time devoted to standards of care, teaching, and testing and the time lost from actually doing "the real work." These professionals frequently complain about the demands of paperwork—recording evidence for insurance companies and government agencies—that attempts to standardize their practice and its evaluation. So, too, factory managers, design engineers, architects, builders, and social workers can easily and quickly compile a lengthy list of codes and regulations that must be accommodated in their work on a daily basis. But the measuring-standardizing activity is often the only thing that people consider "real evidence of results." It is a failure of imagination to believe this.

Modern industrial and urban worlds were built with standards embedded in them. Think of any modern institution: education, the city, policing, the military, the stock exchange. Each is predicated, to some degree, on the tools of measurements, the purchase of standardized commodities (or investments in the futures of commodities) (Cronon 1991), and the measurement and formal presentation of results. And, as thoroughly at the center as these processes are, their mandate often remains unquestioned.

One simple explanation for overlooking the question "Why standardize?" is that standardization is considered to be a necessary technique designed to facilitate other tasks. We often confront standards as fully developed forms, such as an electricity grid or a health regulation. The resulting ahistoricity is another factor that allows the quintessentially sociocultural and ethical aspects of standards to be overlooked. In this sense, the process of standardization is both a hidden and a

---

4. Please note that standardization is *not* exclusive to modernity per se, but it has accelerated with its electronic and global forms, as already described. The distinction between a convention and a standard, or perhaps working standard, is in some instances difficult to make (Michael Evans, personal communication). Over time, what had been social conventions become increasingly standardized in formal ways, after which the difference between a standard and a convention is no longer minor but qualitative.

central feature of modern social and cultural life (Slaton and Abbate 2001). Or, ironically, because standards are so pervasive that they have become taken for granted in our everyday environment, they may become completely embedded in everyday tools of use. (Consider the Japanese toilets that routinely check urine to see if several medical parameters are out of line.) Residual categories (e.g., "None of the above" or "Note elsewhere classified") may help us see the boundaries of standards, for example, rare medical conditions such as being allergic to onions or having undiagnosable chronic pain.

Containing Messy Reality

Looking into the edges and detritus of infrastructure can be a messy and distasteful task. For example, in studying the history of taxidermy, Star found herself tracking down the biological supply houses that had provided items such as standard-size glass eyes for the different animals in the museum dioramas (see figure); homemade devices for shaving and softening animal skins, and other tools for preparing and preserving specimens and habitats. The glass eyes are standardized with respect to color and size, and they are designed to be "lifelike." Because taxidermy sought to take the messy scenes of hunting and capturing specimens and create in their place a clean, almost transcendental vision of nature, more of its craft skills came to require the use of standardized parts and means of working. In this, taxidermy holds common ground with medical illustration; debugging complex computer programs; and moving from the messy, noisy birthing room to a cleaned up, peaceful space.

One final, and rather comical, reason why standards may be neglected in sociocultural research into science and technology is that they are boring. They are often, as mentioned, deeply embedded in infrastructures of various sorts, apparent as wires, plugs, lists, labels, and other semicultural forms. Once these forms are noticed and examined, most social scientists do agree that they are important indeed[5]—but they escape notice, and so studying them can be lonely. Some ten years ago, in Palo Alto, California, several colleagues formed a new professional society. The idea for the society arose from a series of conversations we had about our somewhat unusual research topics—these very semicultural things that most people find quite dull. We called it The Society of People Interested in Boring Things. Among the boring topics that the founders brought to the first meetings were the inscription of gender in unemployment forms used by the city government in Hamburg, Germany; the difficulties of measuring urine output in a post surgical ward in the Netherlands, and how to design better cups for metrication;

5. Including, among others, Charlotte Linde and Susan Anderson anthropologists; Geoffrey Bowker, historian; David Levy, information scientist; Marc Berg, physician/philosopher; Leigh Star; Sigrid Müller; and later Martha Lampland.

PLATE VI.

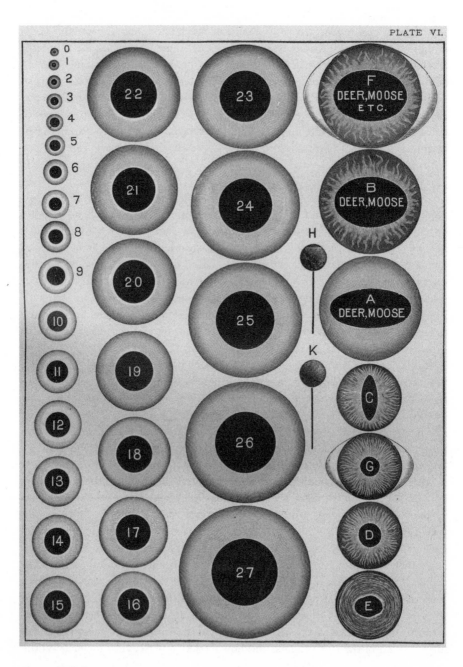

Glass eyeball chart.

From Oliver Davie. 1894. *Methods in the Art of Taxidermy.* Philadelphia: David McKay. Plate VI.

the company mascot and the slogans used by a large midwestern insurance firm in its attempts to build corporate cultures; and how nematologists[6] use computers to keep track of their worm specimens. To put it mildly, these are not central topics in social science—yet!

How should these boring traces that bear evidence of the development of standards be studied? The authors in this book share an ecological sense of the worlds of infrastructure. Collectively, we find it necessary to deconstruct boring backstage elements. In this, we seek to restore the narratives of these standards: their historical development, their political consequences, and the smoke-filled rooms always attached to decisions made about them. This means overcoming the initial boredom of analysis and, furthermore, practicing a deeper ecology than that of input-output or systems analysis. We have to listen to infrastructure and bring imagination to understanding its components and how they work.

Listening to infrastructure does not neatly resolve into accessible questions. The relative weight of social agents and social structure has long posed an analytic dilemma. Various traditions have resolved this tension in different ways. The issue here is less one of relative weighting than of determining when and where we can identify people, objects, and emergent properties (structures) in standardizing and using standards (or not using them). Standards such as human weight charts, blood types, and electrical current now appear fixed and neutral, although this inert quality obscures the enormous amount of work needed to stabilize knowledge, freeze action, delete outliers and residuals, and facilitate use. In the course of studying standardization, the challenge has been to figure out which threads to unravel to make visible and lively the otherwise banal and interstitial character of standards and quantified or formally represented phenomena. We have interrogated stories of the inevitability of technological development and the neutrality of quantification, examined claims about standardized representations and authority, and sought to reveal the political and ethical problems at the heart of these struggles. In other words, when does a structure become a structure and when is a social actor an agent are part and parcel of the story to be told here.

Although standards become crucial (and sometimes visible) once they have been stabilized, the physical features or phenomenal manifestations vary greatly. Enormous infrastructures, such as railroad lines or urban plumbing systems (Collier n.d.), can be massive observable structures of metal and PVC (polyvinyl chloride), whereas other equally central infrastructures, such the computer language ASCII (Pargman and Palme, chap. 7 in this volume) or standards for chronological age (Bowker and Star 1999; Treas, chap. 3 in this volume) or EU environmental policy, are far less tangible. (They are argued through in hardworking committees that rarely see the light of history. At most, they may be manifested in

6. Biologists who study worms, in this case those who were sequencing the genome of the nematode *Caenorhabditis elegans*.

written documents that are rarely perused by any but their most immediate users.) But early, incomplete forms of standardization, too, set parameters within which social action takes place. In retrospect, while doing the "archaeology of things and their order," we may come to see them as clearly the product of a long series of events and actions taken to make them so.

Perhaps the most intriguing aspect of standards is their always already incomplete and inadequate (compared to some ideal) character (for a similar argument, see Barry 2001, 62–84). The push to standardize presumes the ability to constrain a phenomenon within a particular set of dimensions, as well as the ability to dictate behavior to achieve the narrowly defined dimensions that stipulate its outcome. A great deal of work is conducted to make the standard possible, and then this must be followed up by agents committed to implementation and oversight. Again, standardization is a recursive practice, necessarily historical and embedded in a series of complex events and social structures. This is quite evident in legal proceedings adjudicating the application of standards and regulations, but it is not limited to the legal domain.

Formal compliance to standards without substantial change in practice is common. Paperwork is filled out to assure the responsible authorities that regulations have been recognized, but this can be a far cry from actually fulfilling those requirements. Obviously, there is a range of behavior here, from near compliance to outright defiance of regulations. There is also a range of means to reward and punish those who step outside the boundaries of usual (and accepted) practice. For instance, accounting standards—both technical and ethical—were strongly questioned in the period following the ENRON scandal. The unethical behavior could hardly be blamed on the standards themselves. However, large shake-ups and scandals such as this leave a legacy. Frequently, they lead to a rethinking of the shape and usability of the standards. Organizations may begin to seek ways of making standards more effective in rendering an outcome that is consistent across the entire field of use.

The attempt to purify and simplify processes of standardization—through bureaucratic maneuvers or more contested legal procedures—contributes directly to the overdetermined or layered, socially and culturally embedded quality of standards. With time, this process can lead to what Callon (1998) calls "irreversibility." This is, in the first instance, a functional irreversibility—for instance, what would it take to change the meaning of a red light to "go" and a green light to "stop"? Obviously, we would have to invest untold billions, and some sort political platform on which to base the change, to achieve this reversal.

A related maturation and reification process that leads, over time, to complexly recursive standards is that developed by Wimsatt (1998) under the rubric "generative entrenchment." Small changes made early in the life of any developmental system will ramify throughout the growth of the system, becoming increasingly more difficult to eradicate. Wimsatt originally used the example of teratagens, drawing on embryology to illustrate the process. If something small goes wrong early in the development of a fetus, it will ramify systemically. If it happens late in development, it is much more likely to be trivial. So it is with standards. Small conventions adopted early on are both inherited and ramify throughout the system.

Hence, we can say that slippage (such as the made-up phone number in the introductory example) between a standard and its realization in action becomes a crucial unit of analysis for the study of standardization and quantification. Using historical analysis, this may mean analyzing irreversibilities and processes of generative entrenchment. What is being standardized, for what purpose, and with what result? When did it begin? What were its first entrenchments? What can and should be changed? Who are the actors engaged in the process of standardization, and do they change at different moments of a standard's genesis and maturation? What small decisions have ramified through the life and spread of the standards? When does a standard become sufficiently stabilized to be seen as an object or quality influencing social behavior? How do we address the objectlike quality of standards while keeping a keen eye on the necessarily historical and processual quality of its emergence, transformation, and (variably) long life? How do standards developed in one context acquire a modular character, enabling them to be moved around or to serve as templates for the development of other standards?

And as a result of this intensely social, historical process, it is necessary to acknowledge the contingent and, in some cases, arbitrary nature of the standards themselves. The confusion, anger, and frustration people express about standards are easily related to the apparent alogical or irrational character of standards. The association of standards with irrationality demonstrates, as little else can, Max Weber's powerful insight that the move toward modern rationality necessarily resulted in forms of irrationality. The iron cage of bureaucracy has perhaps become a sociotechnical cage—sticky and partly binding but also complexly structured with information architectures and human behavior. This stands in contrast to current neoinstitutionalists' arguments that change proceeds linearly and along traceable tracks.

Types of Standards

What is being standardized, and who is being standardized? What is the difference between a gold standard and a working standard? Moreover, how is the baseline for a standard determined? How does it become naturalized or standardized, so that it slips into the realm of common sense and tacit knowledge? In many cases, the causal relationships will be difficult to parse in any simple fashion. Standardized procedures are created and enforced by governmental agencies; others are created and enforced by private businesses, professions, and local regulations. Still others are created by individual labs, families, and even individuals. We bemoan some standards and laud others; we resist some entirely while gladly imposing others on ourselves.

Another set of issues crucial to the study of standardization concerns the scale and scope of a standard. Standards vary in their scope and scale. The standards for chocolate differ in both scale and scope from standards for gasoline purity. The history of modern standards is one in which the range of standards, and their relative scale and scope, has increased dramatically. This contributes to one of the astounding features of contemporary standardization bodies—the presumption that their work is necessarily global in impact. Clearly, coordinating communication over the World Wide Web or between computers in different sites has necessitated extensive work to ensure ease of data movement and flow, akin perhaps to the grand projects of building railroads and cutting deep river channels to facilitate the movement of goods in the nineteenth century (prompting the use of the metaphor of the superhighway). We must not lose sight, however, of the simple fact that standards are intensely local, in the sense that, despite their global reach, they touch very specific communities in very specific contexts.

In their book *Sorting Things Out,* Bowker and Star (1999) explore the case of the International Classification of Diseases in some detail. It is a good example of several of the issues we have raised so far; and it is, furthermore, widespread, standardized, and old (more than one hundred years old). It thus incorporates legacy systems, multiple (and sometimes competing) architectures, and hundreds of standards. It is clear in this case, also, that Western and middle-class values and foci are inscribed in the list of mortality and morbidity labels. For instance, heroin addiction and absinthe addiction are prominently featured in the drug abuse area of medical classification; petrol sniffing (widespread in the developing world) and legal addictions to pain medicines or Ritalin in the first world are ignored. When we turn to the part of the classification scheme that encodes accidents, a person may fall from an automobile or from a commode (a common accident during the care of elders at home in the developed world) but not, say, from an elephant or a carrying chair. These labels are used, among other things, to fill out death certificates and record epidemics around the globe. They are thus critical, although often invisible, resources for allocating aid and tracking international health concerns, which in turn become standardized and quantified in many ways.

Another crucial feature of standards, related to issues of scale and distribution, is a notion of degrees of delegation. How is the enforcement of standards (and their attendant moral orders) managed? Increasingly in developed states, delegation is managed via home-testing kits, through directives printed out from the pharmacist, or through a wobbly network of social workers and elder-care workers. The importance of degrees of delegation will perhaps help us distinguish between the character of conventions (discussed later), such as the manner in which a doctor treats a patient in a face-to-face encounter, and standards imposed at the state or national level for medical practice, such as provisions for cleanliness or the ways people must dispose of toxic substances. This idea about degrees of delegation bears a resemblance to Bruno Latour's notion of "action at a distance" (1987, 219), but it is also, again, clearly informed by Weber's seminal work on modern bureaucracies and studies of complex organizations.

What Is Infrastructure?

Defining *infrastructure* is not as easy as it may seem. Along the way, we use the term, encounter it as used by others in connection with standardization. We had a commonsense notion of infrastructure when we began discussing the nature of "boring things"—infrastructure is something that other things "run on," things that are substrate to events and movements: railroads, highways, plumbing, electricity, and, more recently, the information superhighway. Good infrastructure is by definition invisible, part of the background for other kinds of work. It is ready-to-hand. This image holds up well enough for most purposes—when we turn on the faucet for a drink of water we use a vast infrastructure of plumbing and water regulation without usually thinking much about it.

However, in light of a deeper analysis of infrastructure, and especially to understand large-scale technical systems in the making or to examine the situations of those who are *not* served by a particular infrastructure, this image is both too shallow and too absolute. For a highway engineer, the tarmac is not infrastructure but a topic of research and development. For the blind person, the graphics programming and standards for the World Wide Web are not helpful supporters of computer use but barriers that must be worked around (Star 1991). To expand on our point about standards, one person's infrastructure is another's brick wall, or in some cases, one person's brick wall is another's object of demolition. As Star and Karen Ruhleder (1996) put it, infrastructure is a fundamentally relational concept, becoming real infrastructure in relation to organized practices (see also Jewett and Kling 1991). So, within a given cultural context, the teacher considers the blackboard as working infrastructure to be integral to giving a lesson. For the school architect, and for the janitor, it is a variable in a spatial planning process or a target for cleaning. "Analytically, infrastructure appears only as a relational property, not as a thing stripped of use" (Star and Ruhleder 1996, 113).

Infrastructure is part of human organization and as problematic as any other. The contributors to this book have done a kind of gestalt switch, what Bowker (1994a) has called an "infrastructural inversion"—foregrounding the truly backstage elements of work practice, the boring embedded things, and, of course, infrastructure. Work in the history of science (Bowker 1994b; Hughes 1983, 1989; Yates 1989; Edwards 1996; Summerton 1994) has begun to describe the history of large-scale systems in precisely this way. In science as well as in culture more generally, we see and name things differently under different infrastructural regimes. Technological developments are processes and relations braided in with thought and work. In the study of nematologists mentioned earlier, Star and Ruhleder listed the properties of infrastructure as embeddedness; transparency; having reach or scope; being learned as part of membership; having links with conventions of practice; embodying standards; being built on an installed base (and its inertia); becoming visible on breakdown; and being fixed in modular increments, not centrally or from an overview.

The strangeness of infrastructure is not the usual sort of anthropological strangeness, in which we enter another culture with a kind of trained suspended judgment, eager to learn the categories of that culture rather than imposing our own. Infrastructural strangeness is an embedded strangeness, a second-order one, that of the forgotten, the background, the frozen in place. It always interacts with any given culture (see, e.g., Akrich 1993 on African use of electricity systems; Verran 2001 on Nigerian uses of mathematics), but it may be both local and global, or multiply standardized and adapted.

The ecology of the distributed high-tech workplace, home, or school is profoundly impacted by this relatively unstudied infrastructure that permeates all its functions. If we study a city and neglect its sewers and power supplies (as many have), we miss essential aspects of distributional justice and planning power (see Latour and Hernant 1999; Collier, n.d.). If we study an information system and neglect its standards, wires and settings, we miss equally essential aspects of aesthetics, justice, and change. Perhaps if we stopped thinking of computers as information highways and began to think of them more modestly as symbol sewers, this realm would open up a bit.

Many aspects of infrastructure are more difficult to locate, for several reasons. First, people tend to discount this aspect of infrastructure as extraneous to knowledge or to their tasks. They, therefore, do not tend to mention them in official historical narratives (except in passing; see Clarke and Fujimura 1992b for an excellent discussion of this problem). Second, details such as materials, standards, and formal modeling assumptions do not always obviously intersect the variables and processes that are familiar to us in analyzing human interactions. The known variables such as gender, race, status, career, power, and innovation trajectories are subtly represented in infrastructures, especially as they appear in processes of standardizing and quantifying (see Stern 2002). Unearthing the narratives behind the boring aspects of infrastructure does, however, reveal (often in a very direct way) how knowledge is constrained, built, and preserved. This book is a modest witness to the bricks of the infrastructure wall that are placed there in the form of codes, protocols, algorithms, and so forth.

Intellectual Background: Science Studies

In the world of science, scholars began to study how laboratories work during the 1970s, work that later linked to these concerns about infrastructure. In Europe and the United States, notably with the publication of Bruno Latour and Steve Woolgar's *Laboratory Life* (1979), people began to explore the laboratory as a kind of anthropological field, with scientists as the tribe. *Laboratory Life* is an ethnographic examination of the production of a scientific fact. It looks at the devices (called "inscription devices" by Latour and Woolgar) used by biologists to record and preserve data. In this, Latour and Woolgar unpack the gradual deletion of uncertainty and qualifications in the statements emerging from the laboratory. They explicitly try to eschew the obvious categories that previous, more macro-scale studies of science produced: occupational stratification, the role of national cultures in science, and so forth. The idea was to approach science afresh, to look empirically at knowledge construction in a detailed, face-to-face context, much as an anthropologist would approach a new tribe (their metaphor).

With the publication of *Laboratory Life*, a window was opened to a more qualitative, intensively observational set of studies of scientific work and practice. Many were produced over the next two decades, examining such interesting phenomena as talk in the laboratory, the acquisition of manual skills in performing tests, the ambiguity of scientific objects, and the intersection of heterogeneous viewpoints in making scientific theories; by the 1990s, the research community began the systematic study of the design and use of information technologies (see, e.g., Star 1995). This development toward the "technical turn" in science studies, that is, the ethnographic study of the design and use of advanced technologies such as computers, had many research ramifications. It used many of the same techniques as the earlier laboratory studies of science; however, it also directly engaged social scientists in studying communicating machines, the emergence of the personal computer (PC) and the World Wide Web, and attempts to model human behavior. In addition, by the early 1990s, several detailed studies of the materials aspects of scientific work began to appear, many of which began to pick up other aspects of boring things, such as the humble stuff used in experiments (see, e.g., Clarke 1998) and the way equipment and its layout reflects a particular scientific commitment.

Recent studies have taken this combination of the technical turn and studies of materials deep into the investigation of infrastructure (see, e.g., Star and Ruhleder 1996). The ethnographic eye that helped reveal the inner workings of science or technology research and development applies no less to the built scientific-technical environment. Arguments about standardization, selection and maintenance of tools, and the right materials for the job of knowledge production have slowly come into center stage via this synthesis (Clarke and Fujimura 1992a). Along with this has come a rediscovery of some of the research tools germane to cognate disciplines that had previously analyzed material culture and the built environment. These include, *inter alia*, fields such as architecture (in which scholars sometimes read the built environment as a kind of text), literary theory (especially those aspects of literary theory that help hidden stylistic assumptions and narrative structure surface), and social geography (in which the values and biases inherent in such tools as maps are a lively topic of inquiry). Work on quantification and standards as structuring knowledge owes much to these fields, as well as to cognitive anthropology and linguistics, areas whose scholars have investigated the toolness and origin of various modeling systems.

An example of the study of a technical project in which infrastructure and standards are central is the sociological study of the biological effort the Worm Community Project of the early 1990s. Leigh Star and Karen Ruhleder (1996) found a world of clashing meanings between designers and users of the system. The project came just before the advent of the Web and as academe become fully saturated with e-mail users (especially in the sciences) in 1991–1994. They studied a scientific community and a custom-made system codesigned with the community. Most respondents said they liked the system, praising its ease of use and its understanding of the problem domain. On the other hand, most did not sign on. Many chose instead to use Gopher and other simpler net utilities with less technical functionality; later, of course, they turned to the World Wide Web. Obviously, this was a problem of some concern to the system developers and evaluators. Despite good user prototype feedback and participation in the system development, there were unforeseen, complex challenges to use involving standards and infrastructural and organizational relationships. The system was neither widely adopted nor did it have a sustained impact on the field as the resources and communication channels it proffered became available through other (often more accessible) means. It did provide insights for social scientists into the profound impact of the understanding of infrastructure on group interactions.

In short, the study showed that problems with local infrastructure and standardization can mean the rise or fall of expensive experiments. Each form of standardizing, quantifying, or modeling stands on top of another, supporting it but not in a smooth or seamless fashion. Some stone walls fall down; some survive for thousands of years. (The same can be said, in interesting ways, of Gothic cathedrals, many of which did fall down; see Turnbull 1993.) Thus, some forms of infrastructure are added to and maintained; some are neglected. In any event, the nesting properties of infrastructure converge with human behavior to form a complexly imbricated, messy whole.[7]

The metaphor of imbrication is important for the rest of this book, in addition to its evocative picture of uncemented things producing a larger whole. Imbrication also implies that each part may shift in character over time as the whole is edited or rearranged. Thus, a keystone at one time—a rigid standard, say—may

---

7. *Imbrication* means partly overlapping layers (not stacks), such as we would find in a good stone fence in New England. As a metaphor, it means the heterogeneous variety of things that partially hold one another up, including discourses, actions, architecture, work, and standards/quantifications/models.

become a minor interchangeable end stone at another, later time. The job of the analyst of scientific or technical work, and its attendant standards, therefore, is to raise these second- and third-order questions about the existence and nature of the whole classification scheme, the taken-for-granted tools used in intra- and inter-disciplinary communication. One aspect of this analysis is to bring to the surface the embedded biases in representations of knowledge, both blatant (e.g., in advertisements) and subtle (e.g., in the categories in databases). "Other" ways of knowing, speaking here in the voice of modern analysts of modern systems of knowing, can become important bridges that reflect back on "our" ways of knowing. Our ethnocentrism, and our assumptions about infrastructure and standards, comes to the fore when we encounter wild (to us) representations. One rich place of encounter, as already noted, is culturally diverse and different kinds of maps. Radically different maps derive from non-Cartesian, relational, cognitive commitments, in which things such time, emotion, and trust often appear explicitly as part of the cartography.

The cultural values in the representation of alternative maps, by contrast, seem fairly transparent, especially in contrast with standardized flat maps. There are underlying standard databases that feed these maps. It is not so easy to access the geographical information systems underpinning many of today's maps, especially those coordinating and standardizing metadata.[8]

Metadata are equally imbued with values, as are all maps, but these values are much harder to pick out. Sometimes this is because they are embedded in numbers or layout; at other times, it is because we rarely get a view of how the metadata are distributed, collected, standardized, or designed (Chrisman 1997). The politics of metadata rarely appears in a way accessible to users. Rather, they are distributed over the bureaucratic, cultural, and military landscapes, appearing as settings, standards, and technical aspects of user's manuals. If we wish to understand more of the deep structure of interdisciplinary communication, it is important to develop good tools for parsing metadata—culturally and politically, as well as technically.

There is much work to be done to understand all the ramifications of this deep approach to standards. We need to understand more, for example, about the behind-the-scenes decisions made about things such as encoding and standardizing, decisions made about tinkering and tailoring activities (see, e.g., Gasser 1986; Trigg and Bødker 1994), and the observation and deconstruction of decisions carried into infrastructural forms. We need to understand more about how metadata develops as well as how it fails to develop, say, in cross-disciplinary work.

8. *Metadata*, a term originating in library and computer sciences, means data about data. Metadata about a library collection, for example, tells us what *types* of documents may be found in a collection (maps, manuscripts, archives, journals, or books) but not the exact titles held by the collection. This is an echo of our introductory example—what happens if someone tries to call our friend at 1-2-3-4-5-6-7? They will encounter a form that is devoid of content.

A deconstructive reading of infrastructure quickly reveals the presence of what literary theorists call a master narrative, that is, a single voice that does not problematize diversity. This is the voice of the unconscious center, the pseudo-inclusive generic. An example of this encoding into infrastructure is a medical history form for women that encodes monogamous traditional heterosexuality as the only class of responses: blanks for "maiden name" and "husband's name," blanks for "form of birth control," but none for other sexual practices that may have medical consequences, and no place at all for partners other than a husband to be called in a medical emergency. Latour (1996) discusses the narrative inscribed in the failed metro system, Aramis, as encoding a particular size of car based on the presumed nuclear family. Band-Aids or mastectomy prostheses labeled "flesh colored," which are closest to the color of white people's skin are another example of an embedded assumption. We may uncover them one by one, as do many of the chapters in this book; however, we also need deeper theoretical analyses to guide our wanderings and also to guide our development of a better way. Millerand and Bowker (chap. 6 in this volume) speak to the double process of deconstruction and working systems juggling real-time as well as archival forms of information. Standardizing and customizing proceed in absolute, messy tandem.

Many information systems represent and encode work processes, directly or indirectly (payroll systems, time sheets, activity reports, and flow charts are among the many infrastructural tools that perform this function in the workplace). Such tools, like language itself, are always incomplete with reference to both the complexity and the indexicality of the processes represented. People are always adjusting, working around standards to get on with their jobs and their lives.

But the solution to these silences and their negative consequences is not always simply making things visible to all. For example, when analyzing the attempts by a group of nurses to classify their work processes, Bowker and Star (1999) see them walk a delicate line between visibility and invisibility. They wanted their work to be represented in order to be legitimated; at the same time, if they categorized all the tasks they did and then built the forms into hospital record-keeping to track that work, they risked having the hospital accountants and health maintenance organization (HMO) officials deskill (see Lampland, chap. 5 in this volume) their work and try to fob parts of it off on less expensive paraprofessionals. So, leave the work tacit, and it fades into the wallpaper (in one respondent's words, "we are thrown in with the price of the room"). Make the work explicit, and it becomes a target for surveillance. The job of the nursing classifiers was to balance somewhere in the middle, making their work just visible enough for legitimation while maintaining an area of discretion.

Much infrastructure is marked with this sort of invisible trouble. In academic departments, the question of what work should be visible and what should count for promotions and tenure often brings this to a head. Researchers who develop large information systems, performing and visual artists, those whose work takes a long time to come to fruition (such as architects) are often at a disadvantage with promotion committees, which may not be able to evaluate or understand the invisible work that goes into research but does not culminate in a book or an article in a refereed journal. Similar problems occur in promotion standards or standards of conduct in large commercial firms.

Boring Things

This introduction offers a short guide to the large terrain of sociotechnical understandings of standards, quantification, and formalization, with an emphasis on standardization. Like all maps, especially those showing relatively unexplored intellectual terrain, it is incomplete, deleting the work involved in making it (although perhaps not wholly, we hope), and has several places where the old "here be dragons" is drawn around the black box of future investigation. Sorting through the richness of things and ideas to create an archive necessarily raises the question of choice and the politics of representation. Not everything can be either known or kept; politics aside, there simply is not room for every piece of paper, artifact, and form of representation. Size limits become political limits: Whose ideas and whose things *matter*? So we reach another kind of one-size dilemma—on the one hand, knowledge has different sizes, metaphorically speaking (and sometimes literally speaking); on the other, the purpose of an archive is to keep stuff in anticipation of the future, and it is hard to know beforehand what will be useful. The imperative to know is paired with the ability to keep and to hold; authority arises from classification as well as from ownership. Within the computer and information sciences, there is a serious utopian dream of remembering everything equally (for instance, there is a project begun years ago by computer scientist Douglas Lenat, the Cyc project, to store all commonsense knowledge into a huge electronic encyclopedia). These visionary musings of ever-expanding storage space obscure (one more time) the politics of collection and memory (Bowker 2006). These politics are irrevocably central to constructing archival projects, even large ones such as the Cyc project. Common sense shifts as mercurially as language does; retrieval questions are still organized by algorithm, paid-for space in an information field, and other questions of social stratification. The famous search engine Google, like all commercial search engines, sells electronic real estate allowing a firm's name to come up first in a search, even though other hits appear further down in the listing. Battles over whose knowledge will be remembered and who has rights to remember it will be fought visibly in conferences and computer centers, but will ultimately reside in the structures of data themselves, including their political, commercial, and sponsorship attachments. As Lampland (chap. 5 in this volume) shows, larger-scale political events and structures influence how work is remembered, how attempts to standardize and remember it may change radically over time, and how different forms of knowing and exchange (peasant measurements and bartering systems vs. centralized attempts to measure work hours) may be at war with one another.

One of the areas barely explicated is the difference between standards and conventions of human behavior. We name as standardized many examples of rote, repeated behavior devised according to a script or proscription. We messily and intentionally, therefore, traipse into the part of the map claimed by many sociologists and anthropologists as norms of behavior, conventional types of action, or the sorts of standard actions developed around material constraints and the functions of social worlds, such as described by Becker (1982) in his fertile analysis of these processes, *Art Worlds*. Why is a play usually two to three hours long? Becker argues that this timing derives from an intersection of constraints on work. The finances of production and wages for actors, security guards, parking attendants, and fast food conveyers are such that breaking this timetable may become too expensive. Over time, this convention becomes widespread—although never absolute.

Behavioral and technical norms also influence infrastructural elements such as time of performance. Babysitters must be paid so that parents can attend a play and must be home in time to go to school at a set hour (perhaps even a standardized hour when the bells and lockout system become imbricated with standard protocols for computer systems). People train their bodies to sit for so long and no longer. In the West, the tolerance for silences, confusion, and multiple voices is variable but, as a rule, not huge.

Where do such conventions and norms become standards, quantities, or parts of formal models? As we may expect, there is a leaky border among all of these sorts of action and inscription. We have rules of thumb,[9] not written guidelines, for everyday life. They meet conventions in a loose conglomeration of quantified inscriptions, technical delegation, and actions both locally and at a distance (standards), but do not usually include transient customs such as skirt length, habitual turns of phrase, or locally specific times when meals are usually eaten.

As with all conclusions, this is clearly not a satisfactory or comprehensive exploration of how these sorts of things meet. In fact, on this view, the rule of thumb we have described is *both* conventional and standard. This book lives in the middle and offers a number of empirical examples and analytic concepts that may help us, if not to clean up the messy imbrications, at least enjoy them, understand how they work, and minimize the suffering that overly specific or underly specific approaches may produce.

---

9. *Rule of thumb* is itself a fascinating term, dating back to a legal term and a social system still, sadly, in full swing; in early modern times it referred to the thickness of the implement with which a husband may beat a disobedient wife. Naturally, thumbs differ, but none is as big as a baseball bat or a log for burning in the fire.

■ Standards Where You Least Expect Them

MARTHA LAMPLAND

In the process of compiling this book, I thought I had acquired a healthy respect for the imperative to standardize in modern society. I was wrong. I laughed when I heard that federal prosecutors intended to enter into evidence an alleged application for Islamic warrior training at Jose Padilla's trial. Did this look anything like an application for summer camp? I was chastened, however, when I learned of the bureaucratic fastidiousness of Al Qaeda's leaders.

Padilla Case Changed a Lot in 5 Years

The Trial of the Onetime "Dirty Bomb" Suspect
Is Opening; An Alleged Jihad Form May Be Key

CAROL J. WILLIAMS

Los Angeles Times (Home Edition)   May 13, 2007

When federal prosecutors begin to present evidence Monday against terrorism suspect Jose Padilla, their case is expected to rest heavily on a single document: his alleged application to become an Islamic warrior.

The federal indictment says Padilla filled out the mujahedin data form on July 24, 2000, "in preparation for violent jihad training in Afghanistan." The indictment alleges Padilla and two codefendants sought U.S. recruits and funding for foreign holy wars.

Prosecutors plan to call a covert CIA operative to testify in disguise about the document's provenance and chain of possession, and will go on to introduce more than half of the 200-plus transcripts from wiretapped conversations among the defendants.

Penalty for Crossing an Al Qaeda Boss?

A Nasty Memo

Sebastian Rotella, Los Angeles
Times Staff Writer

Los Angeles Times   April 16, 2008

Mohammed Atef was furious.

The Al Qaeda leader had learned that a subordinate had broken the rules repeatedly. So he did his duty as the feared military chief of a global terror network: He fired off a nasty memo.

In two pages mixing flowery religious terms with itemized complaints, the Egyptian boss accused the militant of misappropriating cash, a car, sick leave, research papers and an air conditioner during "an austerity situation" for the network. He demanded a detailed letter of explanation.

"I was very upset by what you did," Atef wrote. "I obtained 75,000 rupees for you and your family's trip to Egypt. I learned that you did not submit the voucher to the accountant, and that you made reservations for 40,000 rupees and kept the remainder claiming you have a right to do so."

The memo by Atef, who later died in the U.S.-led assault on Osama bin Laden's Afghan refuge in 2001, is among recently declassified documents that reveal a little-known side of the network. Although Al Qaeda has endured thanks to a loose and flexible structure, its internal culture has nonetheless been surprisingly bureaucratic and persistently fractious, investigators and experts say.

The documents were captured in Afghanistan and Iraq and date from the early 1990s to the present. They depict an organization obsessed with paperwork and penny-pinching and afflicted with a damaging propensity for feuds.

They scrupulously documented logistical details—one memo accounts for a mislaid Kalashnikov rifle and 125 rounds of ammunition.

■ Separate, Together

ALICE DREGER

Wall Street Journal (Eastern Edition)   New York, N.Y. July 9, 2003

Who isn't saddened to hear that Laleh and Ladan Bijani—the twins from
Iran, joined at the head—died from the surgery they hoped would separate
them? Only 29 years old, intelligent, educated, engaging, basically
healthy—here were two women who put their faith in God and medicine,
trusting that "Operation Hope" would allow them the physical independence they craved. Still, it is hard to account for the level of interest and
subsequent grief shown by perfect strangers around the world. What was it
about this tragic tale of two women born conjoined that resonated with so
many people?

For some, stories of separation surgeries like the story of the Bijani sisters
seem to be stories of earthly justice. In the classic version, two innocent
babes, victimized by a whim of nature, are liberated by heroic surgeons,
freed to live the lives they were always meant to have. This is medicine—
indeed, humanity—at its best.

For others, separation surgeries like the Bijanis' constitute tales of the
near-universal struggle to achieve individuality, integrity, and something
vaguely known as freedom. For still others, separation surgeries become
narratives of national identity. Perhaps because these cases are so often
international in nature they necessarily become arenas of national politics;
the Bijanis' was even labeled in the style of an international military
campaign.

Americans take pride in the fact that many of our best surgeons pack
their tools and trek all over the globe, exporting the values of Lady Liberty
by providing extraordinary care to the weak and downtrodden. Meanwhile,
many of the Bijani sisters' compatriots saw the women's unrelenting quest
as proof of both Iranian piety and Iranian modernism. Compare this to the
situation, in September 2000, when British surgeons sought out the courts'
permission to effectively kill one conjoined twin to save her sister over the

wishes of the girls' Maltese, Roman Catholic parents. During that long and painful saga, many Brits, particularly the justices involved, saw the need to establish secular, modernist (British) values over what they took to be primitive, even extremist (Maltese), religiosity.

The funny thing is, most people who are conjoined don't think conjoinment is cause for international quests or battles, and that's because they don't think being conjoined is all that bad. In this sense, the Bijani sisters were truly unusual. Clearly Laleh and Ladan felt unduly confined and intolerably limited by their conjoinment. But if you look at the history of what people who are conjoined have said about themselves—and what they have decided about separation for themselves—it appears most people who grow up conjoined see conjoinment as a necessary, integral and acceptable part of their individuality. Some have even confessed they consider it a superior state in its provision of perpetual companionship.

Lori and Reba Schappell of Pennsylvania, conjoined at one side of the head, have said precisely what dozens of their historical predecessors did: We are not broken, not in need of repair; [w]e are normal people, even if we look very different from others physically. Even as young children Abigail and Brittany Hensel (who are now entering their teen years conjoined) found others' questions about their state rather dense and tiresome; they, like their parents, understood that this is just who they are and that conjoinment need not unduly limit each girl's experience of human individuality. Some people born conjoined have used their anomalies as a means of gaining freedoms otherwise unavailable to them. So Eng and Chang Bunker, the 19th-century conjoined brothers who coined the term "Siamese Twins" to refer to themselves, took advantage of their state to travel the world and earn fairly sizable fortunes. Far from experiencing their conjoinment as a form of entrapment, Eng and Chang understood that their unusual bodies gave them something of a leg up in the world.

Notably, few conjoined people old enough to decide for themselves have pursued the question of separation, and the Bijani sisters appear to have been the only adult conjoined twins in history to go through with such an operation. Even when their twin was dying, many conjoined twins have accepted their own deaths over separation. The sisters known as the Biddenden Maids, born in the year 1100, lived until the age of 34, and when one died the survivor refused offers of separation, declaring, "As we came together we will go together." The second sister died about six hours after the first. Eng Bunker does not appear to have explicitly asked for separation when his brother Chang died, even though according to the biographers Irving and Amy Wallace, Eng "had continued rational" after becoming aware of Chang's death. When told by his son that Chang was dead, Eng is reported to have declared, "Then I am going!" And in 1967, after learning that Margaret Gibb, the sister to whom she was conjoined, had advanced cancer, Mary Gibb did not seek separation but instead opted to face the cancer and subsequent death together with her sister.

To those of us born physically separate from all but our mothers, it may seem a paradox to consider conjoinment as part of one's individuality. But it isn't that hard to understand if you realize two things. First, not unlike people who are conjoined, many of us singletons find ourselves in intimate human relations which seem to augment rather than to confine our individuality. For example, though some people find parenthood and marriage unbearably confining, many of us experience these relationships as limiting, yes, but also as necessary, integral, and quite acceptable.

Second, not unlike people who are conjoined, most of us can't imagine living in a body other than the one with which we were born. Even though I know, as a woman, that I'm subject to peculiar forms of discrimination (just as people who are conjoined are subject to peculiar forms of discrimination), I can't imagine being otherwise. It's who I am. Statistically speaking, it is fairly unusual for a person to feel that she was born into a body which unjustly imprisons her. Most people, including most people born with socially challenging anatomies, accept their congenital states. That includes most people born with dwarfism, with intersex (mixed sex anatomy), or conjoined. It also includes most people who were born male, most born female, most born white, and most born black.

That said, it does seem reasonable to let transgendered people, people with short limbs due to achondroplasia dwarfism, and people like the Bijani sisters—people who feel they were born into a body that is intolerable— seek out surgeons willing to change them. But even as those quests happen, and especially because it looks like they are happening more and more, we ought to be asking to what degree these people feel they must change their bodies because others around them will not change their minds.

To what degree did the Bijani sisters have their individuality limited not by their conjoinment, but by others' perception of their conjoinment? If history is any guide, chances are they were told unequivocally that they could not marry, could not have children, could not be full persons while they remained conjoined. Chances are that, like gay men and lesbians in the U.S. until a few days ago, like many girls and boys of all races throughout our history, they got the message that their anatomies must necessarily limit their political and social identities.

What Laleh and Ladan went through was an extreme example of the human quest for individuality. Even as we try to make sense of their deaths, we ought not to forget that that quest takes place in a world of others to whom we are metaphorically bound, others who have the chance to limit us or to free us in critical, though non-surgical, ways.

Ms. Dreger, an historian of anatomy at Michigan State University, is the author of "One of Us: Conjoined Twins and the Future of Normal," forthcoming from Harvard University Press in the Spring of 2004.

Photo taken at Ellis Island National Monument by Susan Leigh Star.

■ Ellis Island

SUSAN LEIGH STAR

For the millions of immigrants coming into New York's Ellis Island during the twentieth century, *standard* and *substandard* took a quick and brutal form. Inspectors at the gates spent about eight seconds per person determining if the person was pregnant, had a communicable disease (especially tuberculosis), was a "degenerate" of some form, was literate, and, in short, whether he or she should be allowed into the United States. The information management technology was simple: white chalk was used to make a mark on the clothing of the person involved. The figure on the facing page is an image of the medical metrication system.

BEYOND THE STANDARD HUMAN?

Steven Epstein

■ "One Size Definitely Does Not Fit All," declared the headline for the lead article (Duenwald 2003) of the *New York Times*'s annual special section on women's health, beneath a photo of a physically and ethnically diverse group of women, all dressed in variations on the theme of white shirts and khaki pants. The article was about women's physical dimensions, but with that title it might have been about anything related to health, medicine, or the body—indeed, a year earlier, another story by the same reporter described criticism of universalistic approaches to hormone therapy under a nearly identical headline (Duenwald 2002). Repudiations of one-size-fits-all thinking have become ubiquitous in the arena of health and medicine in the United States, and they are asserted with a vehemence that invites scrutiny.

In this essay, through particular attention to recent policy debates focused on the use of human subjects in biomedical research in the United States, I examine attempts by what might be called an antistandardization resistance movement to displace the standard human. I argue that medical research has been subject to critique especially because researchers have been thought to make use of a typological standard human—human subjects drawn from a particular population subtype, in this case adult white men. Although a critique of the standard human research subject

This chapter stems from my participation in a residential research fellowship program at the University of California's Humanities Research Institute. I am grateful to David Theo Goldberg and the HRI staff for their assistance, and to the members of the group—organizers Leigh Star and Martha Lampland, along with Geof Bowker, Rogers Hall, Jean Lave, Martin Lengwiler, Janice Neri, Ted Porter, Mimi Saunders, and Judith Treas—for comradeship, productive intellectual exchange, and comments on previous drafts. Martha Lampland and Leigh Star deserve special credit for all their efforts in producing this volume.

might have led either to attempts to construct a single, more representative standard or, at the opposite extreme, to an insistence on radical particularity, in this case the solution has been what I call niche standardization—the development of multiple standard subtypes distinguished by sex, race, ethnicity, and age. Thus, this critique of standardization in biomedical research has led not to less standardization of the human but, rather, to a new standardization that operates according to different principles.[1]

Standardization and Resistance

Biomedicine is hardly the only social domain that has developed conceptions of the standard human (see also, Martha Lampland, chap. 5; Judith Treas, chap. 3 in this volume). Many disparate actors have sought to standardize the human for different purposes—economic, political, administrative, and practical (Scott 1998; Bowker and Star 1999)—and modern life would be unimaginably cumbersome if the much-vaunted uniqueness of the individual were everywhere acknowledged. Yet, not infrequently, these standardization projects have become flashpoints for resistance, for campaigns of opposition waged behind banners of fairness, equity, and justice. We tolerate or invite standardization of many things, from consumer goods to bureaucratic policies to technical codes (Bowker and Star 1999). But we are often troubled by the notion that there is or might be a standard human.

Attempts to construct a standard human are unavoidable, in part because other standards have spillover effects. To standardize consumer goods is inevitably to standardize those who consume them; to standardize policies is to standardize those administered by them (Busch 2000). When McDonalds hamburgers are made uniform, their makers assume the uniformity of the tastes and metabolic processes of the people who eat them; there is little space for recognition of those who are, say, allergic to onions (Star 1991). Or consider the seats on airplanes. It seems sensible and cost-effective to make them all the same size within compartments, except insofar as that presumes that all people can adjust to the same seat dimensions and seat pitch. The brouhaha in 2002 over a new Southwest Airlines policy is instructive. When the company announced that ticket agents would begin requiring those who could not fit into one seat to purchase the adjoining seat as well, it invited a wave of protest and ridicule. A *Rocky Mountain News* editorial

1. I conducted the research for this chapter as part of a larger research project (Epstein 2003, 2004, 2005, 2007). Data for the larger project were obtained from seventy-two semi-structured, in-person interviews; documents and reports from federal health agencies and the U.S. Congress; archival materials from health advocacy organizations; materials from pharmaceutical companies and their trade organizations; articles, letters, editorials, and news reports published in medical, scientific, and public health journals; and articles, editorials, letters, and reports appearing in the mass media.

predicted that beleaguered passengers, already forced to shed shoes and belts at the security checkpoint, would soon be subject to the final humiliation: "they'll tape-measure your behind" (Sharkey 2002, C9).

The Southwest episode was a public relations stumble for an airline that cultivates a cheery, populist image. But the outcry against the new policy demonstrated the conviction—call it deep-seated—that there is something fundamentally unjust about expecting that all humans will conform to a standard type. And if this example seems frivolous, it is not hard to think of cases in which the costs to the nonstandard among us are far more severe than the price of an extra plane ticket or a bit of public embarrassment. In the design of air bags, for example, assumptions about the standard human have meant serious injury and death for children and smaller adults crushed by the deployment of devices intended to save lives. A decision by the U.S. Department of Transportation to correct this health hazard by reducing the force of air bags to a new standard level was predicted to result in a trade-off that would "reduce the risk to children and small adults, possibly at the expense of men who drive without seatbelts" (Wald 2000, A-14). The marketing of pharmaceutical drugs provides another telling example. A standard dosage is generally prescribed for all adults, despite research findings on the wide range of individual variation in responsiveness to medications and despite the consequent risk of adverse drug reactions (Cohen 2001, 20). "One dose fits all is a marketing myth," observes Carl Peck, the former head of the Center for Drug Evaluation and Research of the Food and Drug Administration (FDA), "but it's the holy grail that every drug company tries to achieve" (2001, 26).

In this essay, I focus on the case of pharmaceutical drug development and clinical research more generally. In this domain—where knowledge about the few must be generalized to the many—the debate over the standard human is posed in especially stark and interesting terms. Medical experiments necessarily are conducted on a restricted group of individuals—human subjects recruited to participate in controlled trials designed to study biological processes or test the safety and efficacy of medical interventions. Findings from those trials are then extrapolated to the human population as a whole. To the extent that the participants' experiences, as studied in the trial, are representative of those of humanity broadly, then the experimental findings are trustworthy. But what if the composition of clinical trials differs from the overall population in detectable ways? For example, what if, as critics began claiming in the late 1980s, adult white men were overrepresented as research subjects while women, people of color, children, and the elderly were underrepresented? What does this mean for the reliability of biomedical knowledge about those various populations? And what does it say about our society if one particular subset of humanity—the adult white male—is, in effect, taken to be the standard human type?

After all, it is one thing to define a standard human in *statistical* terms—to say that the standard human is the modal type or that the standard human is the mean, plus or minus a standard deviation or two. This, in effect, is the logic of the aircraft designer who determines that most of the bell curve of human behinds can be squeezed into 17-inch seats with a tolerable level of discomfort. It is another thing when notions of the standard human are *typological*. In cases of this sort, the determination of standards seem consciously or unconsciously to single out a particular sociodemographic group as the ideal specimen of humanity—the ones most worthy of study—and then to treat knowledge derived from the study of this group as universal truth, applicable without modification to all other groups in society. As I describe here, those who found U.S. biomedical research guilty of using a typological standard believed that what was at stake was not just the credibility of medical knowledge but social justice in a strong sense.

My analysis proceeds in three steps. First, I consider some of the pathways by which notions of the standard human may have come to be adopted within biomedicine; then, I examine the critiques mounted against this standardizing practice; and, finally, I describe the new forms of standardization that have been implemented as innovative solutions in the face of political and scientific challenge.

L'Homme Moyen and the Golden Mean

Modern attempts to standardize humanity are linked closely with the development of the sciences of statistics and probability. When Adolphe Quetelet announced the new science of social physics in 1831, its central concept was what he called *l'homme moyen*, the average man. This man would have not just an average height, weight, education, and length of life but also an average propensity to marry, commit suicide, or engage in criminal acts (Gigerenzer et al. 1989, 41–42). As Gerd Gigerenzer and coauthors note, Quetelet fully understood that *l'homme moyen* was an abstraction who existed nowhere in reality: "But abstraction was essential to social science. Real individuals were too numerous and diverse for psychological study to contribute much to an understanding of the social condition" (Gigerenzer et al. 1989, 41).

L'homme moyen prefigured present-day preoccupations with the standard human in several respects. First, and important, Quetelet's conception of the average man was not merely statistical but also normative: "Quetelet had always idealized the mean as the point of virtue lying between vicious extremes, and the embodiment of moderation in a world forever threatened by revolution. . . . Actual individuals are merely imperfect copies of the virtuous golden mean" (Gigerenzer et al. 1989, 54). *L'homme moyen* was "normal" in the double sense; he was the midpoint of natural variation but also the way that people were supposed to be. Thus, this definition of the standard human served as an instrument of normalization as the term has been used by Foucault (1979)—a way of defining norms, but simultaneously a way of monitoring and measuring deviation from those norms. Nowadays, notions of the standard human are invoked not only to shore up biological and cultural norms of all sorts but also, quite often, to justify the correction of such deviations. For example, in July 2003, the FDA approved the licensing of human growth hormone shots for children who are "idiopathically short"—that is, children whose shortness is not linked to any medical condition such as hormonal deficiency but who simply happen to find themselves at the low end of the normal distribution of heights (Food and Drug Administration 2003).[2]

A second way in which Quetelet's work anticipated some of today's debates lay in his interest in the gradual evolution of the mean. The qualities of *l'homme moyen* were not invariant, Quetelet understood, and thus a study of those qualities over time provided clear statistical evidence of the trajectory of a society (Gigerenzer et al. 1989, 42). Today's demographers continue to concern themselves with the social and political significance of shifts in the averages—the declining age of first intercourse, the declining number of children, and so on. The consequences of such developments can be far-reaching, as the recent changes in American body shapes make clear. "Many people in this country no longer fit in the standard-size casket," a funeral industry executive recently commented in the *New York Times*. "The standard-size casket is meant to go in the standard-size vault, and the standard-size vault is meant to go into the standard-size cemetery plot. Everyone in the industry is aware of the problem" (St. John 2003, A1). Moreover, a reliance on obsolete depictions of the standard human can be injurious to human welfare. In 2003, the Federal Aviation Administration ordered airlines to increase their assumed average passenger weight by 10 pounds—in recognition of the bulging of U.S. waistlines, but, more immediately, in response to an airplane crash that may have resulted from a fatal underestimation of the passenger load (Wald 2003).

Medicine: From the Principle of Specificity to Standardization

It was by no means foreordained that conceptions of the standard human developed by social scientists such as Quetelet would find their way into medicine. Rather than dealing in universals, medicine has often been preoccupied with differences, both between individuals and between social groups. Indeed, nineteenth-century U.S. physicians were committed to the principle of specificity—the belief that medical therapy "was to be sensitively gauged not to a disease entity but to such distinctive features of the patient as age, gender, ethnicity, socioeconomic position, and moral status, and to attributes of place like climate, topography, and population density" (Warner 1986, 58). As John Harley Warner describes, "Admonitions to heed the various elements encompassed by the principle of specificity permeated therapeutic instruction. . . . Professors routinely taught that

2. On the long history of transforming variations from the norm into medical conditions warranting treatment, see Conrad and Schneider (1980); Conrad (2005).

[these various] individuating factors . . . modified the character of the disease and the operation of drugs . . ." (Warner 1986, 58–59). Clearly, from the standpoint of this medical worldview, *l'homme moyen* simply had no place; physicians literally would not have known how to treat him. Only in the latter part of the nineteenth century, with the increasing adoption of European theories of scientific medicine, did U.S. physicians gradually abandon the notion that treatments should be tailored to the idiosyncratic constitutions of patients in favor of the idea that each specific kind of illness required a distinctive treatment that might be applied universally to sufferers (Warner 1986, 248–49).

In the twentieth century, a host of developments converged to encourage a relentless (if often resisted) standardization of medical practice: the rise of modern methods of pharmaceutical drug testing and drug regulation, the development of epidemiological studies based on notions of statistical risk, the codification of international classification systems for morbidity and mortality, the increased reliance on standard protocols and expert systems, the rise of evidence-based medicine as a social movement within biomedicine, and the growth of managed care as a system of rationing and surveillance (Berg 1997; Marks 1997; Rothstein 2003; Timmermans and Berg 2003). All these developments privileged a conception of medicine as a science (consistent, predictable, and transparent) over a competing conception of medicine as an art (dependent on intuition, experience, and embodied skills, and respectful of the particularities of individual patients) (Gordon 1988).

The standardization of medical practice has often carried with it a strong presumption that the object of medical attention—the patient—could likewise be conceived of in standard and universal terms. It is telling that the actors who are trained to simulate disease symptoms for students in the medical school classroom are referred to as "standardized patients." Now used in all medical schools in the United States to provide students with real-life experience (Garfinkel 2001), standardized patients are labeled as such because the actors endeavor to present a consistent simulation each time they perform (Wallace 1997). Mechanical simulators of patients—such as the virtual-reality surgical simulators currently used in some medical teaching settings—carry this standardization to an even greater degree, helping thereby to standardize the practices being taught by means of the simulators (Johnson 2005, 146).

With the standardization of the patient, biomedical notions of the standard human are also exported from the clinical setting to inform—and coercively structure—many mundane aspects of the lives of the healthy. Gina Kolata describes the consolidation of the standard that a person's number of heartbeats per minute during an exercise workout should never exceed 220 minus the person's age. The standard is derived from a study of "volunteers who were likely not representative of the population but were 'whoever came in the door.' " (Kolata 2001, D1) Moreover, the numerical finding from the study was simply an average. But "the formula became increasingly entrenched, used to make graphs that are posted on the walls of health clubs and in cardiology treadmill rooms, prescribed in information for heart patients and inscribed in textbooks" (D1). In a nice example of the often hidden, prescriptive power of standards that have been built into everyday technologies (Bowker and Star 1999), some exercise machines are actually designed to shut down automatically if a user exceeds 90 percent of the supposed maximum heart rate (Kolata 2001).

The Human Subject as Standardized Work Object

With the standardization of medical practice and the standardization of the patient came attempts to construct a new standardized object for biomedical research—the human subject. As science studies analysts have shown more generally, the production of generalizable knowledge from laboratory settings invariably presumes the creation of standardized "working objects" (Daston and Galison 1992, 85) whose essential characteristics can be claimed to vary little from one laboratory to the next (Latour 1987; Clarke and Fujimura 1992a; Porter 1995, 29–32; Jordan and Lynch 1998). "No science can do without such standardized working objects," note Lorraine Daston and Peter Galison, "for unrefined natural objects are too quirkily particular to cooperate in generalizations and comparisons" (1992, 85). Although such objects are often inanimate, such as scientific instruments and procedures, they may sometimes be living things—for example, the fruit fly, whose natural variability had to be reduced before it could become a reliable experimental object (Kohler 1994). However, creating a standard human for research purposes is potentially more problematic. Whereas scientists could construct a variety of *Drosophila* for laboratory purposes, clinical researchers studying humans are obliged to take those humans essentially as they find them. Researchers seeking a standard human "working object" therefore have several options: they may assume that differences between humans are irrelevant for their purposes; they may seek out only those individuals who have chosen characteristics; or they may subject different subpopulations of humans to separate or comparative studies.

Of course, researchers may seek to bypass this dilemma by substituting experimental animals drawn from other mammalian species, such as rats or chimpanzees, in place of the standard human. But this solution, although common in many biomedical contexts, may engender even more controversy over the generalizability of the findings (Löwy and Gaudillière 1998). First, it is by no means obvious that other animals are any more easily standardized than humans, and attempts to employ them in medical testing have sometimes run afoul of significant intraspecies variation (Porter 1995, 29). Second, there may often be considerable uncertainty about the implications for *Homo sapiens* of findings in other species. Indeed, a culture that has been growing suspicious of extrapolating findings from men to women, adults to children, or white people to people of color may have little patience with the notion that "rats are miniature people" (Busch, Tanaka, and Gunter 2000, 108; see also Clarke 1995). Moreover, as work by Nik Brown and Mike Michael suggests, there is a built-in tension in the reliance on animals models, stemming from a precarious simultaneity of perceived similarities and differences—to the extent that a species *is* seen as sufficiently close to humans for scientific purposes, then the moral legitimacy of manipulating them for our ends may become increasingly suspect (Brown and Michael 2001).

If animals cannot solve our problems by freeing us from reliance on human experimentation, then perhaps, some imagine, we can bypass humans in a different way: by *simulating* a trial using computer models. Companies specializing in the design of clinical trial simulation software hold out the hope of taking data from preliminary trials on the absorption of drugs by body tissues and feeding these data into computer models, which can then project the effects of therapies on so-called "virtual patients." Indeed, virtual patients could (in theory) be designed who "match expected real patients in terms of distribution of age, weight, gender, disease severity and so on" (Pollack 1998, D1). But this research is in infancy, and it is never expected to substitute fully for conducting clinical trials in living people.

False Universalism?

Restricting our attention, then, to nonvirtual members of our own species, we can ask: Who has come to serve as the standard human in biomedical domains? As I have already suggested, a conventional wisdom that took shape in the late 1980s and that solidified in the 1990s points to the overrepresentation in clinical research of the adult white male, tacitly taken to be a stand-in for the population as a whole. Bernadine Healy, the first female director of the National Institutes of Health (NIH) in the early 1990s, subsequently expressed this view as follows:

> [T]he orthodoxy of sameness and the orthodoxy of the mean, which has dominated much of the thinking in medical science . . . often impaired our attitude toward clinical research in those days—we tended to want to reduce the human to that 60 kilogram white male,[3] 35 years of age, and make that the normative standard—and have everything extrapolated from that tidy, neat mean, "the average American male" (Science Meets Reality 2003).

Critics suggested that this narrow conception of the standard human had thoroughly penetrated medical theory, practice, education, and training, and that its signs ranged from the anatomical images used in medical textbooks (Lawrence and

---

3. Healy may have misspoken here, or there may be an error in the transcript. Medical references are usually to "the 70-kilogram male"—or more recently, reflecting an oft-bemoaned tendency toward amplification in *l'homme moyen*, "the 75-kilogram male."

Bendixen 1992) to the presumptions about which sorts of people could best work as doctors or scientists (Jensvold, Hamilton, and Mackey 1994).[4]

From the late 1980s onward, the demand for change away from this model was propelled by a diverse set of individuals and groups, including politicians, health advocates, interest groups, and social movements (Epstein 2004, 2007). By and large, women's health was the driving wedge. Because of the relative successes of liberal feminism and the rise of women into positions of prominence in government and in science, the question of the appropriate inclusion of women attracted by far the most public attention; concern with other groups followed in its wake. In this regard, the key actors were female members of Congress and their staff (especially the Congressional Caucus for Women's Issues); female insiders at agencies such as the NIH and the FDA; health professionals and scientists sympathetic to women's health issues; activist lawyers; the broader women's health movement; activists concerned with the effects of diseases such as AIDS and breast cancer on women; and new, professionalized women's health advocacy groups that arose in the early 1990s, such as the Society for the Advancement of Women's Health Research (Auerbach and Figert 1995; Narrigan et al. 1997; Weisman 1998, 2000; Baird 1999). But other actors also played important roles in expanding the critique, including pediatricians and geriatricians, politicians concerned about minority health issues (including members of the Congressional Black Caucus), physicians working primarily with racial and ethnic minority patients (including groups such as the National Medical Association, an African American physicians group), and pharmaceutical companies interested in diversifying their markets or developing niche markets.

Those who, in the late 1980s and early 1990s, began calling for reform attributed the use of adult white men as the standard human to various factors, including gender bias on the part of male researchers and a reliance on certain biomedical work routines, such as the use of Veterans Administration patients as samples of convenience for some medical studies. In the case of the underrepresentation of racial and ethnic minorities, many white researchers complained that they were hard to recruit—especially African Americans, who were said to reject the role of medical guinea pig out of suspicion due to the long history of medical experimentation on black people that dates to slavery and includes the infamous Tuskegee study of "untreated syphilis in the Negro male" (Corbie-Smith 1999).

In addition, analysts claimed that underrepresentation was an unintended and unfortunate consequence of government protectionism. In the 1960s, with the publication of shocking reports of widespread abuses of patients in U.S. medical

4. Of course, biomedical institutions were hardly unique in such tendencies, and they took their cues from the broader society. For an interesting historical example of the everyday privileging of whiteness as a normative characteristic of bodies, see Alexandra Stern's description of "better baby" contests in Indiana, which "promoted the idea that only White babies could achieve perfection and symbolize the Hoosier state" (2002, 748).

experiments, policymakers had asserted that stricter measures were needed to safeguard human subjects (Rothman 1991, 70–84). This wave of concern culminated in the enactment of formal legal protection of the rights of experimental subjects. A distinguishing feature of this regime of regulation was its emphasis on the protection from harm of vulnerable populations: children, fetuses, prisoners, the poor, and the mentally infirm (National Commission 1979). However, advocates of inclusion pointed to the irrationalities that stemmed from well-intentioned protectionism—for example, that 75 percent of the drugs that physicians were prescribing to children had never actually been tested in clinical trials with pediatric populations (Yaffe 1983). Advocates also decried the drastic unintended consequences of protectionism for women's health. As a result of a 1977 FDA rule, women of childbearing potential were formally excluded from many drug trials, whether they were pregnant or not, or had any intention of becoming so, out of concern that an experimental drug might bring harm to a fetus (as, for example, the drug thalidomide had done to women sometime earlier, especially in Europe).

Many anecdotal reports also attributed the underrepresentation of women to biomedical researchers' belief that women were "complicated" research subjects; their monthly fluctuations in hormone levels could confound the effects of the medical regimes or therapies under investigation. Men's bodies, by this reasoning, were simpler to study, because there were fewer "variables" to control for (Mastroianni, Faden, and Federman 1994, 80). Somewhat similarly, researchers presumed not just that elderly patients deserved protection from the risks of biomedical experiments but also that the tendency for the elderly to suffer from multiple health problems and consume many medications simultaneously would muddy the results from clinical trials. Most generally, some researchers believed that trials with a so-called homogeneous subject population constituted a quicker route to medical knowledge, whereas the noise of variation would introduce heterogeneity that would inevitably complicate the interpretation of findings and make them more difficult to publish in medical journals. Left unstated, of course, were the presumptions about who constituted "noise." In these various ways, as Rebecca Dresser observes, "NIH officials and biomedical researchers have, consciously or unconsciously, defined the [adult] white male as the normal, representative human being. From this perspective, the goal of advancing human health can be achieved by studying the white male human model" (Dresser 1992, 28).

In pressing their arguments about why a false standard was problematic, reformers placed particular emphasis on the argument that underrepresentation matters because of medical differences across groups—differences generally understood in biological terms. Although such claims in some ways harked back to the nineteenth-century principle of specificity, reformers positioned themselves as representing the latest in medical knowledge. In recent decades, biomedical findings of differences, particularly ones relevant to pharmaceutical drug development and use, have multiplied. For example, many studies have analyzed the unequal distribution by sex and race/ethnicity of the cytochrome P450 enzymes that metabolize many pharmaceutical drugs, and researchers have argued that such differences cause a standard dose of these medications to have different effects in different groups (Lin et al. 1991; Levy 1993). Researchers also became aware of significant differences in the efficacy of antihypertensive drugs across racial groups in the United States, and consequently the idea that African American patients respond better to diuretics than to other drugs such as beta-blockers rapidly became an established medical maxim (Moser 1995).[5] Pediatricians, insisting that a child is not simply a miniature adult, have also put forward difference arguments, noting that not just children's size but also the maturational process itself affects the course of disease and the physiological responsiveness to medications (Yaffe 1980). Analogously, geriatricians have contended that age-associated physiological impairments and reduced rates of drug clearance make it more likely that the elderly will suffer from adverse drug reactions (Schmucker 1984).

To be sure, many clinical researchers responded that medical differences between groups were relatively trivial in comparison to the fundamental similarities across the human species. Moreover, at least in the case of racial and ethnic differences, some ethicists and scientists expressed concerns that assertions of biological differences might inadvertently trigger a resurgence of scientific racism (King 1992; Schwartz 2001). For the most part, however, those who pointed to biological differences appeared to be relatively unbothered by the risk—well suggested by historical studies of biology and medicine (Gilman 1985; Proctor 1988; Duster 1990; Schiebinger 1993)—that difference might be conceived of in pejorative terms and used to bolster arguments about social inferiority. Instead, reformers used uncertainties about human sameness and difference to their strategic advantage. After all, if the medical differences between, say, women and men were insignificant, then there was no basis to the practice of excluding women so as to reduce variation and simplify the experiment. And, conversely, if the differences mattered, then it was particularly crucial to avoid extrapolation and to study both groups; any failure to do so reflected clear bias. In either case, the remedy was the same—the inclusion of more women (and, by implication, other groups) in research populations.

Representational Complexities

It is beyond the scope of this chapter to provide a full assessment of these various arguments about inequities in the conceptualization of the standard human in biomedical research. But it bears observing that the claim that the heterosexual,

---

5. In recent years, the question of whether there are meaningful racial differences in the effects of antihypertensive drugs has become more controversial. For one recent summary, see Seventh Report (2003).

middle-age, adult white male has served as the biomedical standard is simply too sweeping when posed in universal terms.[6] Several issues and examples are relevant to thinking about the variegated histories of standardizing the human in the biomedical domain.

First, given the paucity of past data on the demographics of research participants, and the wide range of ways in which *underrepresentation* might be defined or measured (Mastroianni, Faden, and Federman 1994, 30), the issue of whether women or other groups *were* underrepresented has remained somewhat controversial (Bird 1994; Mastroianni, Faden, and Federman 1994, 47–68; Meinert and Gilpin 2001). In the absence of any systematic record-keeping or reporting of the demographics of research subjects by the NIH, the FDA, or medical journals, a range of empirical studies of research demographics, using different samples and methodologies, have arrived at opposing conclusions (although a majority of them have pointed to past histories of underrepresentation to some degree). For example, many analysts have emphasized the greater numbers of men in studies that recruited both sexes, especially for certain conditions such as heart disease and HIV/AIDS. But others have pointed to the higher number of studies of conditions exclusively or primarily affecting women than for conditions exclusively or primarily affecting men. Thus (as examples discussed later also suggest), although much of the criticism of research practices is warranted, the broad-brush claim that adult white men had become the universal human subject fails to do justice to the varying particularities of research designs.

Second, an overrepresentation of white men as research subjects in many studies may not necessarily mean that white men have been unthinkingly taken as the standard and that knowledge gleaned from studying them was simply extrapolated to others. It may, in fact, mean that researchers simply cared most about white men, whom they understood to be importantly distinctive. Nancy Krieger and Elizabeth Fee, noting that strongly held beliefs about differences have long been a constituent feature of modern medical and epidemiological research in the United States, argue that to imagine that medical researchers took white men as the norm (and hence representative of other groups) is to misread how medicine privileged white men as a distinct group:

> [B]y the time that researchers began to standardize methods for clinical and epidemiological research, notions of difference were so firmly embedded that whites and nonwhites, women and men, were rarely studied together. Moreover, most researchers and physicians were interested only in the health status of whites and, in the case of women, only in their reproductive health. They therefore used white men as the research subjects of choice for all health conditions other than women's reproductive health and paid attention to the health status of nonwhites only to measure

---

6. I discuss these issues in more detail in Epstein (2007, chap. 2).

degrees of racial difference. For the most part, the health of women and men of color and the nonreproductive health of white women was simply ignored. It is critical to read these omissions as evidence of a logic of difference rather than as an assumption of similarity. (Krieger and Fee 1996, 21)

This interesting and important critique raises questions about whether or when it makes sense to speak of a biomedical reliance on a standard human.

Third, clearly a wide range of circumstantial factors often have determined which subjects have been recruited to which studies. Insofar as many researchers do take "whoever walks in the door," researchers may tend to oversample those who, for mundane and important reasons (access to child care, access to transportation, or ability to take time off from work) find themselves more available to participate. But samples of convenience may be defined in even more specific ways, as Heather Munro Prescott reminds us in an article aptly titled "Using the Student Body" (2002). As Prescott notes, "the assumption that undergraduates are natural research subjects is so deeply embedded in both the history of and present-day thinking on human experimentation that it is difficult to separate discussion of student subjects from that of other healthy volunteers" (2002, 5). Studies of students have been used to establish a range of baseline physiological measures and standards of normality, including "the normal ranges for blood pressure, lung capacity, pulse rate, basal metabolism, and other physiological processes" (16). For these purposes, students have been considered "ideal" in the double sense: a handy "captive population," but also ideal specimens of human normality (37–38).

Yet, fourth, at least sometimes, the availability of captive populations may result in standards being constructed with reference to people whom researchers otherwise might not consider to be ideal specimens. The historian Londa Schiebinger has provided an illuminating look at how medical experimenters selected their subjects in eighteenth-century Europe. She notes, "To some extent, the choice of subjects was simply arbitrary. As with dissection, physicians and surgeons used any bodies they could lay their hands on (perhaps legally and morally, perhaps not)" (Schiebinger 2004, 386). Prisoners, hospital patients, orphans, and soldiers were among those most likely to be experimented on, and "from a medical point of view, there was nothing special about these bodies, except their availability" (393). Work by other historians likewise has suggested how the socially vulnerable sometimes have come to serve not only as the experimental subject of choice but also, as a consequence, as the standard human. The use, in the antebellum South, of black American corpses as subjects for anatomical dissection is a case in point. At a time when medical institutions resorted to contracting with grave-robbers to obtain corpses, black bodies were simply more vulnerable to expropriation. But the result is that nineteenth-century physicians developed ideas about human anatomy based largely on the study of African Americans (Blakely and Harrington 1997). As Robert L. Blakely and Judith M. Harrington observe, "It is one of the ironies of medical history that, although blacks were generally regarded as 'inferior' or even 'subhuman', their corpses were considered 'good enough' to use in the instruction of human anatomy" (1997, 163).

Lisa Cartwright offers an interesting example of how the availability of human materials may result in unusual choices of research subject in her article "The Visible Man: The Male Criminal Subject as Biomedical Norm" (1997). Cartwright tells the story of Joseph Paul Jernigan, an executed murderer whose "body was posthumously selected from numerous cadavers circulating in the public domain to serve as the raw material for an anatomical compendium of the male body that is currently available on the World Wide Web, in CD-ROM, and on video as the Visible Man" (1997, 125). As a white male of a certain size and weight who was executed while in good health and whose body was in the public domain, Jernigan was perceived as a desirable choice for the project. Cartwright may be exaggerating when she suggests that, as a result of this selection, "Jernigan's body has become the gold standard of human anatomy" (125). But she underscores the irony that Jernigan's availability to serve as an abstract and universal subject depended precisely on his radical otherness and particularity, as constituted in juridical terms (136).[7]

A final example of the arbitrary factors that may sometimes determine who comes to serve as the standard human was provided in 2002 by Craig Venter, the director of the private initiative to map the human genome. When Venter "came out of the closet" as the primary source of the genetic material studied by his company, a *New York Times* headline expressed the relationship between scientific truth and personal identity as "Scientist Reveals Genome Secret: It's Him" (Wade 2002).[8]

By cataloguing these examples, my point is not to doubt the observation that the adult white male has frequently been taken as the standard human type in medical thinking and practice. But I do mean to suggest the variation and subtlety lurking beneath an apparent uniformity. A more adequate understanding of the diverse imaginings of the biomedical subject awaits a good deal more scholarly investigation.

Remaking the Standard Human: Responses to Perceptions of Illegitimacy

Broadly speaking, the complaint that a designated standard human has been falsely treated as representative of the species has been addressed through three distinct strategies of reform. First, a common strategy has been to retain the goal of a single

---

7. It should be noted that there was also a Visible Woman, and that together they made up the Visible Human Project. For more on the project, see Waldby (2000).

8. In a sense, Venter was following in a long line of medical researchers who have served as their own experimental subjects (Altman 1987).

standard but to seek ways of making the standard more representative (and, hence, more legitimate) by basing it on greater numbers and more diverse sorts of people. For example, in describing the problematic origins of the maximum heart rate for exercise, Kolata notes the recent efforts of researchers to come up with a more dependable formula. On the basis of published studies encompassing 18,712 people, and drawing on their own research with 514 research subjects, exercise researchers have promulgated a new standard. Instead of 220 minus a person's age, the maximum heart rate is now thought to be 208 minus 0.7 times a person's age (Kolata 2001). Somewhat similarly, pediatricians and statisticians threw out the established standardized growth charts for infants and children that were based on studies of approximately 10,000 children from Ohio. The charts were replaced with new ones, based on a considerably more diverse data set drawn from national health and nutrition surveys of millions of children (Markel 2002).

A second strategy for addressing the lack of representativeness of the standard human is far more radical. Taking furthest the slogan that "one size does not fit all," it repudiates standardization altogether and embraces the particularity of the individual. Although it might be thought that such an extreme resistance to generalization and abstraction would find little support in biomedicine, this move does in fact resonate with certain tendencies in genomics. A new field—pharmacogenomics—has adopted the goal of correlating the discovery of genetic variation with the discovery of differences in drug responsiveness. Front-page articles in the *New York Times* and *Washington Post* have heralded the pharmacogenomic revolution, suggesting that within a few years our doctors will be giving us genetic tests before prescribing medications to determine which particular drugs in which doses are most likely to be of maximum efficacy and minimum toxicity, given our individual genetic profiles (Kolata 1999; Weiss 2000). "Knowing what's in your genes could also take some of the routine guesswork out of medicine," was how *Time* magazine expressed the promise of these techniques (Gorman 1999, 78). It could also go a good distance toward reducing so-called adverse drug reactions, many of which may be caused by the reliance of drug companies on one-size-fits-all dosing.[9] This goal of individualizing therapy is of interest to the business community (Robbins-Roth 1998), and it also has become a mantra at the FDA, where experts speak of individualization as the "third great age of drug development," following on previous successes in addressing first safety and then efficacy (Robert J. Temple, interview, Rockville, Md., August 11, 2000). In 2000, the National Institute of General Medical Science (one of the institutes that make

9. Jay Cohen (2001) describes how FDA review procedures may incline drug companies to use high doses in order to generate sufficiently convincing evidence of the efficacy of their products, but how these doses then cause problems for many patients that become widely apparent only after a drug has been marketed. For a perceptive analysis of the problem of adverse drug reactions, see Corrigan (2002).

up the NIH) awarded $12.8 million in grants to establish a Pharmacogenetic Research Network—a national consortium of research groups working to create a public database of genetic variation in response to therapies.[10]

At this juncture, however, the notion of individualized therapy remains a slogan rather than anything approaching reality (Hedgecoe 2004). Many commentators in the worlds of both research and industry have noted that, "from the industry perspective, the subdivision of a market into smaller markets is hardly ideal" and have questioned whether pharmaceutical companies will find it profitable to develop and market drugs at the level of genotypic difference (Rothstein and Epps 2001, 229). Ultimately, it may prove to be the case that pharmacogenomics encourages not so much an individualization of therapy as a classification of therapy based on broader categories of group membership, such as sex, race, and ethnicity. This is because many of the variations in drug response or disease expression that have been detected at the genetic level have been found to correlate roughly with those categories rather than varying in endlessly individual ways (Foster 2003). In this sense, pharmacogenomics may prove to be less an example of rejecting universalism and standardization altogether and more an example of the next strategy.

The third strategy, which I find especially interesting, is what I call niche standardization (Epstein 2007). Here the goal is neither to develop the perfect new standard that can be universally applicable nor to repudiate standardization altogether in favor of purely individualistic approaches. Instead, researchers and policymakers develop a set of distinct standards, each appropriate to a subgroup: one standard for men, another for women; one standard for blacks, another for whites, another for Asians; one standard for children, another for adults; and so on. Instead of a standard human, we have an intersecting set of standard subtypes.

Niche standardization has become so much a part of the cultural logic of societies such as the United States that it can be found at work not only in obvious places, such as marketing and opinion polling, but in all sorts of unlikely locales. Consider, for example, the crash-test dummy used by automobile manufacturers and safety experts to predict the outcomes on human life of car crashes. How should the crash-test dummy be designed? By constructing a composite of many sorts of automobile drivers and passengers, we might seek to develop the single, perfect crash-test dummy that comes closest to representing the effects of collisions on the average person—the standard dummy. Alternatively, those who doubt the utility of the standard dummy to model the real-world car-crash experiences of distinct individuals might favor the development of fully individualized dummies—but, of course, it would be thoroughly impractical to expect companies to repeat their tests using millions of different dummies. The compromise proved to be niche standardization—the production of a few standard types of

10. See the NIH grants website, http://grants1.nih.gov/grants/guide/rfa-files/RFA-GM-99-004 .html.

dummies: male, female, child, and infant (Fox 2005). Another interesting case concerns attempts to map the size of the average American body. The first such survey since the Second World War was completed in 2004 under the sponsorship of clothing and textile companies, the army, the navy, and several universities. After measuring 10,000 adults using a light-pulsing 3D scanner, the investigators redefined the average American, who not only had grown heavier in recent decades but also could best be comprehended as a series of subtypes. The average American was either a male or a female; either white, black, Hispanic, or "other"; and either 18–25, 36–45, or 56–65 years old. Thus, the survey presented twenty-four standard subtypes, each with average chest, waist, hips, and collar measurements (Zernike 2004). In these examples and others, the trick is to determine which particular ways of differing matter for the problem at hand—typically, sex, age, and race—and then to institutionalize that typology as the standard.

If the existence of a single standard human was the presumption of Fordism (the early-twentieth-century system of producing mass commodities as standardized as the Model T), then niche standardization moves us squarely into the world of post-Fordist manufacturing—niche marketing of diverse products to well-defined subgroups. In a new drug regulatory environment that is sensitive to arguments about group differences, some pharmaceutical companies unable to establish the universal efficacy of their products have, when possible, sought FDA approval to market their products only for use in a designated subgroup. In 2002, Zelnorm, a drug made by Novartis to treat irritable bowel syndrome, became the first of two drugs so far approved by the FDA for use only in women for a condition that affects both sexes (Pollack 2002). The drug's advantages over a placebo in a mixed-sex population could not be demonstrated with statistical significance. In 2005 another drug, called BiDil, which treats heart failure, became the first drug ever approved for use only in African Americans (Kahn 2004; Saul 2005).

Niche standardization also has been a primary consequence of legislative and bureaucratic reforms devised to address the underrepresentation of women, racial and ethnic minorities, children, and the elderly in randomized clinical trials. In response to the fierce complaints against underrepresentation that I have described, Congress, the NIH, the FDA, and other agencies within the Department of Health and Human Services have enacted a series of measures designed to encourage or require both the inclusion of various populations in biomedical research and drug development and the measurement of differences across those groups (Epstein 2004, 2005, 2007). For example, the NIH Revitalization Act, signed into law by President Bill Clinton in 1993, requires not only that women and minorities be included in NIH-funded clinical research but also that clinical trials be "designed and carried out in a manner sufficient to provide for a valid analysis of whether the variables being studied in the trial affect women or members of minority groups, as the case may be, differently than other subjects in the trial" (National Institutes of Health 1993). Both the NIH and the FDA now insist that researchers tabulate human subjects according to sex, race, ethnicity, and age categories and report biomedical or pharmacological differences for these subpopulations.

Niche standardization of this sort has at least potentially displaced the standard human in favor of multiple standards. But in the process, it has ushered in a new regime of standard operating procedures for biomedical research—that is, more standardization, although of a somewhat different sort. The reliance on niche standardization also replaces the question of who comes to serve as the standard with new standardization dilemmas: How should the subcategories themselves be standardized? (The NIH and the FDA rely on census categories to give precise meaning to *race* and *ethnicity*—despite the fact that these categories have no clear biomedical significance and that they shift with each census in response to political pressure; Wright 1994; Nobles 2000.) And which particular ways of differing will be the ones that count in the first place? For example, in recent years, lesbian and gay health advocates have sought to institutionalize a concern with their health issues at the federal level by arguing that sexual orientation is a form of difference analogous to gender and race, and that this variable should be recognized in federal policy documents and health surveys, as well as in clinical trials (Epstein 2003). Such developments suggest a range of difficult questions about the extent to which we deem it advisable to translate the mobilization categories of identity politics into the research categories of biomedical science. When and where does an insistence on biomedical difference truly benefit the constituencies who organize on that platform (Epstein 2007)?

Ironies of Resistance

I have sought to describe and analyze the biopolitical context within which the assertion that "one size does not fit all" has achieved such ubiquity and significance in relation to questions of health and the body. The issues raised by opponents of reliance on a typological standard human in medical research—issues of social justice and the equal right to be healthy and safe—are crucially important ones, even if these critics have oversimplified the history of who has served as the representative research subject. In different ways, the three strategies I have described here—constructing a single, more inclusive and representative standard; embracing particularity; and developing niche standardization—might serve as effective remedies, even though each has its drawbacks in terms of justice, cost, and efficiency. But it is important to say that none of the three seems likely to move us beyond the standardization of the human, if that should be deemed the goal.

Nor can any standardization strategy eliminate the uncertainty that is endemic to the clinical research enterprise. As experts on the techniques of the randomized clinical trial aver, no methodology (short of studying everyone) ultimately guarantees the external validity of trial findings for the mass of humanity or for any specific subset of it. This is true, in part, because the subjects are not laboratory rats but human beings, whose compliance with the experimental protocol cannot always be easily monitored or ensured. Thus clinical trials are often messy, and the true significance of their findings may be subject to intense debate (Epstein 1996, 1997).[11] But, in addition, clinical trials do not employ random samples. Whatever its sociodemographic composition, a clinical trial is populated by individuals who are among the subset of patients who happen to meet the complex, medically defined inclusion/exclusion criteria for the study in question *and* who then agree to participate; such people, by definition, are not representative of the population of patients in question (Meinert 1995, 795). Given that fact, generalizing from the results of any clinical trial is, in the words of one expert, "always going to be a risky business" (Curtis L. Meinert, interview, Baltimore, Md., March 24, 1998).

The problem of uncertainty is equally well demonstrated by new, advanced, differentiated, crash-test dummies seen (no doubt rightly) as an improvement over the family of dummies that served as their predecessors. How are the newest dummies made? Their physical properties are designed on the basis of computer models. The computer programs that generate these models rely on results from an understandably little-publicized research practice know as "cadaveric testing" (Brown 1999). Thus, our belief that experiments on advanced crash-test dummies tell us what will happen when metal meets flesh rests on our faith in the strength of a long representational chain that encodes a series of (generally hidden) assumptions—that corpses respond similarly to living people; that the properties of corpses can be captured in computer models; and that, once modeled, those properties can be transferred seamlessly to engineered objects. In a pragmatic world, we may have no choice but to make such assumptions. But we would do well to remind ourselves, at least on occasion, that the assumptions are there. Neither invoking a standard human nor adopting any of the critical responses that have evolved out of objections to such invocations will rescue us from the uncertainties inherent in making claims about what the mass of humanity is like.

11. Here I leave to one side the question of research biases that may reflect the political economy of pharmaceutical drug development (Tokarsi 2003).

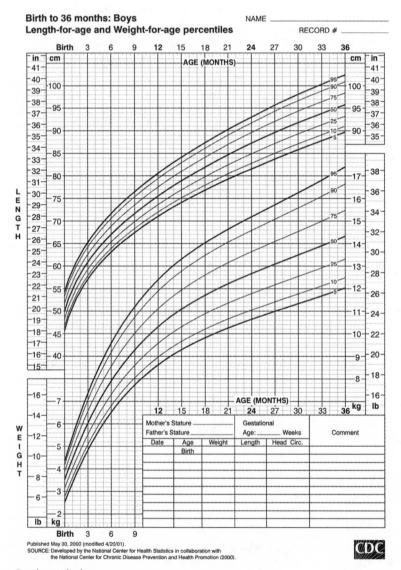

Birth to 36 months: Boys
Length-for-age and Weight-for-age percentiles

NAME _____

RECORD # _____

Boys' growth chart.

Developed by the National Center for Health Statistics in collaboration with the National Center for Chronic Disease Prevention and Health Promotion (2000). Available at: http://www.cdc.gov/growthcharts.

■ The Range of Growth Parameters for Healthy Infants

In the original publication, the boys' chart is in blue and the girls' chart is in pink.

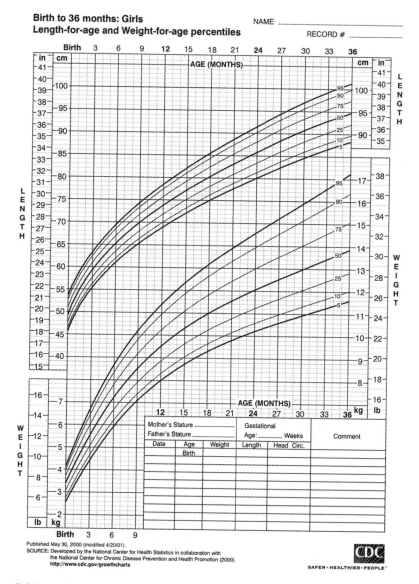

Girls' growth chart.

Developed by the National Center for Health Statistics in collaboration with the National Center for Chronic Disease Prevention and Health Promotion (2000). Available at: http://www.cdc.gov/growthcharts.

■ Portion Creep

Standards and Supersizing

JUDITH TREAS

Americans are fat. In 1999, 61 percent of adults were overweight or even obese, according to the body mass index (BMI), the common weight-to-height measure (National Center for Chronic Disease 2003). One caloric culprit is portion creep, the gradual increase in the amount of food consumed at one time (Young and Nestle 2002).

Do not look to packaged foods or restaurant servings as a guide to reasonable portion size. Food purveyors use bloated portions as a selling point for everything from soft drinks to frozen diet dinners (Schlosser 2001). Standard product sizes are calibrated to showcase bigness. The smallest coffee advertised at Starbucks is a "tall"; the "grande" beverage may be had for only a few cents more. Over time, portions have gotten bigger and, thus, more fattening. In the 1950s, MacDonald's French fries came in only one size; today, that size is labeled "small" and dwarfed by heavily promoted super-sizes that are three times bigger.

Accustomed to Big Gulps and Hungry Man entrees, Americans have lost sight of how much to eat. In the battle against portion bloat, we confront conflicting advice on healthy and virtuous eating. Many diets and weight-loss programs offer their own advice. The government has rung in with two federal serving-size standards. Individuals watching their weight can choose between the food pyramid and the nutrition facts standard.

Most of us remember the U.S. Department of Agriculture *Food Guide Pyramid* from our elementary school classroom. The updated pyramid translates science-based nutritional standards into a few broad food groups to guide food consumption. Today, it advises us to load up on six to eleven servings of breads and pastas from the pyramid's base, but to limit ourselves to two to three servings from the meat, fish, dry beans, eggs, and nuts group at the top (U.S. Department of Agriculture 2000). Caloric requirements determine exactly how many servings to eat. Teenage boys, who need 2,800 calories to get through a day, must eat more servings than children, ages 2 to 6, who can manage on 1,600 calories.

The pyramid takes into account typical portion size reported in food consumption surveys as well as some other factors. Nutrient content is critical for some foods. The servings for eggs, nuts, and beans are set to provide the same amount of protein and minerals as one ounce of meat. For cooked fruits and vegetables, however, a serving is one-half cup, regardless of the particular food's nutritional content, because this makes it easier to use the food pyramid as a consumption guide. Last, one slice constitutes a serving of bread because this is the historical standard from earlier food guides. The serving sizes of other grains are pegged to the calories in this one slice.

Confusingly, labels on package foods use a different standard (U.S. Department of Agriculture 2000). Following the Food and Drug Administration (FDA) *Nutrition Facts* standard, the label gives the standard serving size, the number of servings per package, and other nutritional information. The small print is apt to show that a single muffin or juice carton contains several servings. For products such as cheesecake that are customarily divided into servings, a serving is a fraction of the whole (e.g., one-eighth of the cake). When foods come in discrete units (e.g., crackers), a number closest to the "reference amount" constitutes a serving. This reference corresponds to the amount of the food usually eaten at one time, according to national food consumption surveys. In the case of bulk products such as cereal, a serving is defined in terms of the common (i.e., standard) household measurement (e.g., cup or tablespoon) that is closest to the "reference amount."

If you want to compare the calories in an English muffin and a bagel, either standard works fine. For many foods, however, a serving from the pyramid is not the same as a serving based on the nutrition facts standard. The one-cup serving of cooked pasta described on a box of tortellini equals *two* one-half-cup servings from the food guide pyramid. Although these different federal standards may create confusion, the real public health threat comes from the marketplace, where heavily promoted super-sizes are no longer tethered to reasonable portion size.

Note: There is a nice picture of a bagel today and twenty years ago at http://hin.nhlbi.nih.gov/portion/portion.cgi?action=question&number=1.

■ A Bulge in Misses 8?

Digital Scanners Resize America

MARCIA BIEDERMAN

The New York Times   November 27, 2003
Correction Appended Section G, Circuits
(Late Edition—Final)

Ann Ulrich knows that her days as a model could end soon. A national survey has just finished taking electronic measurements of a cross section of Americans, the apparel industry is eagerly awaiting the results, and she may find that she's no longer their perfect size 1X.

"When you're no longer the closest fit, you are replaced," said Ms. Ulrich, who doesn't need much makeup for her work. Apparel companies hire her for garment fittings because, like other women she represents at Ann Ulrich Fit Model Agency in Minneapolis, she meets the exact measurements for a standard clothing size.

But the buzz at her Midwestern agency—and in the fashion buyers' offices of New York and the garment factories of Hong Kong—is that the current specifications for every size, from 1X to women's 8 to men's medium, may soon be sent out for alterations.

In September, the last of 10,800 volunteers stepped into a booth at a shopping mall, stripped nearly naked and submitted to a 10-second body scan under white light. In exchange for $20 in cash, a $25 gift certificate or, in some cases, no giveaway item, the volunteers contributed to a database that could give manufacturers vital marketing information while helping shoppers find clothes that fit, either in stores or online.

Scanners used in the study traveled the nation for a year, stopping at malls and campuses in 13 cities. The $1 million survey, SizeUSA, is a project of the Textile Clothing Technology Corporation, or TC2, an industry research group that developed the scanners. The scanning project was sponsored by the Commerce Department and private companies, including large retailers that make clothing carrying their labels.

"Forty-one sizes of bra traveled with us, and 11 sizes of shorts," said Jim Lovejoy, the project director. For an earlier national survey in Britain, TC2 scanned women in their own bras, but found that light, silky ones played havoc with the white light. So American women were issued regulation gray cotton. Men in boxers had to change into briefs.

The technology maps the human image as thousands of data points. As subjects stand grasping handles, the scanner registers every detail from the underarm to the crotch, Mr. Lovejoy said, which are among the body areas that matter most in ensuring a proper fit. Software developed by TC2 extracts dozens of measurements per individual from the data.

Donna Lehmann, one of several hundred people who were scanned for SizeUSA at the Fashion Institute of Technology in Manhattan, said the experience "was really cool because it was so futuristic and high-tech." She recalled standing on a platform and hearing a whir as her image was captured.

Ms. Lehmann, 32, the fashion institute's Web coordinator, said she and others scanned there chose to view their images (although the majority of SizeUSA participants did not, Mr. Lovejoy said). "Even for the skinniest girls, it was very disconcerting," said Ms. Lehmann, a size 8. "No one looks good in 3-D." Still, she said she was glad she took part "because sizing is just so ridiculous."

Next month SizeUSA will begin selling a report that correlates the volunteers' measurements with their income, age, gender, ethnicity and ZIP code, the clothing size they believe they wear, and other data. The report costs $20,000, but customers can use a query system to buy a few answers to specific questions for several hundred dollars. Most buyers are likely to be apparel companies, but there could also be interest from the aviation and automotive industries, which collect their own data, and makers of furniture and medical equipment.

Mr. Lovejoy would not disclose preliminary results, but he did drop one bombshell: of 4,000 women included in the survey, less than 10 percent met the definition of misses size 8 established by ASTM International, formerly the American Society for Testing and Materials, which develops standards through industry consensus for products, systems, and services. Yet the fit of today's clothing is often still based on the notion that size 8 is closest to the midpoint of the size range of American women. Some companies, sensing that the standards are obsolete, have already moved the base line up to size 10. The sizing survey aims to substitute data for hunches.

"The basic message will probably be that we're changing shape," Mr. Lovejoy said. The nation's growing obesity has already been well noted, he said, but the report will yield actual measurements, breaking down the data into market segments. He said the data seemed to indicate that Americans have gotten taller as well as heavier, but weight is increasing faster than height, with the result that body proportions no longer fit size specifications set decades ago.

What's more, although the ASTM standards for women's sizes were refined in 1988 and 1990, they are based on data collected in 1941 and do not reflect the nation's ethnic diversity. Efforts were made to recruit minority participants for the recent survey, resulting in a sample that was about 46 percent nonwhite.

The results could yield more predictable fits, provided that apparel companies abandon the growing practice of "vanity sizing," or purposely assigning lower numbers to sizes in order to flatter, say, a size 12 woman who finds she fits into size 10 pants.

The sizing survey could also revive interest in the original purpose for which the scanning technology was developed: to speed the manufacture and lower the price of custom-tailored clothing, thereby limiting returns and markdowns.

Five years ago the industry threw its support behind the research with the idea of bringing custom tailoring to the masses: in addition to taking measurements, TC2's technology extracts data that can be used to make patterns. Now, retailers are more cautious. In the United States, Mr. Lovejoy said, retailers were excited by TC2's plans for mass customization until the downturn in the dot-com technology sector.

At its flagship Manhattan store, Brooks Brothers still takes measurements for custom tailoring with a scanning booth that attracted wide notice when installed in 2001. Two years later, that scanner remains the lone one in the Brooks Brothers chain. Although the digital tailor turns out custom suits faster and more cheaply—suits made in 15 days start at $700—it accounts for only 30 percent of custom business at the store. Many customers still prefer "the personal attention of the tailor," said Joe

Dixon, the company's vice president for production and technical services.

Another drawback with the scanner, Mr. Dixon said, is that it is "very large."

Recently TC2 began production of a streamlined scanner, the fourth generation of a device that measured about 21 by 15 feet by 8 feet high when first produced in 1997—a daunting size for retailers concerned about selling space. At $40,000, compared with $125,000 for the first model, the new scanner is 5 by 9 by 8 feet.

BenchMark Clothiers, a two-year-old company in Little Rock, Ark., has ambitious plans for marketing scanners made by TC2 and taking the technology a step further into custom manufacturing.

Formerly a wardrobe consulting business, BenchMark is seeking deals with men's clothing stores to install scanners and provide custom garments. The company has been testing its system at a store it opened near Little Rock, charging less than $500 for most custom suits and $40 to $80 for a custom shirt. Clothes are ready in two to three weeks.

"At this point, the technology is expensive to implement," said Kevin Lewis, president of BenchMark, adding that his company has invested hundreds of thousands of dollars in the effort. Consumers can select from three types of fit—loose, standard or athletic. In addition, BenchMark has programmed the system to categorize a customer as one of five different body types.

The company is also working on technology that would funnel scanned measurements directly into a computer-assisted-design program that cuts the pattern. That would eliminate the need for a master tailor to translate the data into a pattern.

Information gleaned from the survey could eventually send a new wave of fit models to mannequin makers for scanning. But Eddie Lemack, the sales director of Alvanon, a mannequin company in Manhattan, noted that before long, the fit models may no longer have to travel far. Their digital twins can simply be shipped.

CORRECTION-DATE: December 12, 2003
CORRECTION: An article in Circuits on Nov. 27 about the use of computer scanners in a survey to develop new standard apparel sizes misstated the percentage of women who met the definition of misses' size 8,

and it described the criterion incompletely. The survey found that fewer than 10 percent of those with a size 8 bust, not 10 percent of all the women surveyed, fitted the existing size 8 standard—that is, corresponded in all three measurements (bust, waist and hips).

■ Investment Allocation by Age

JUDITH TREAS

The financial services industry relies on chronological age for a rule-of-thumb to guide consumers' choice of asset allocation strategy. Age categories—assumed to capture biographical detail affecting risk tolerance and the immediacy of need for investment income—determine the recommended mix of "safe" income investments and "riskier" growth investments.

Investment allocation by age (%)

	20–39	40–59	60–69	70–79	80+
Bonds	0	20	40	70	80
Growth and income funds	55	45	35	20	10
Mid-cap funds	15	15	10	0	0
Small-cap funds	15	10	5	5	5
International funds	15	10	10	5	5
Total	100	100	100	100	100

Source: http://www.bankrate.com/brm/news/dollardiva/20000615a.asp. Accessed January 12, 2008.

AGE IN STANDARDS AND STANDARDS FOR AGE

Institutionalizing Chronological Age as Biographical Necessity

Judith Treas

■ How old are you? Even before toddlers are out of diapers, we take for granted that they will be able to answer this simple question. In fact, teachers regard youngsters who arrive at preschool without having mastered this bit of autobiographical information to be woefully unprepared for school and for life. And although it is enough if preschoolers can give their age in years, older children must also know their birthdays. This information permits them to report their age with great numerical precision. As comedian George Carlin (2008) observed, "If you're less than 10 years old, you're so excited about aging that you think in fractions. How old are you? I'm four and a half!" We accept that chronological age is an important marker of social competence and an essential tool for navigating the life course. Certainly, our society makes age—precise chronological age—something that no one should be without.

In focusing on chronological age, that is, time since birth, my objective is to consider how this bit of personal biography came to be used to create a widely employed standard for segmenting and stratifying individuals. Incorporated into the standards imposed from above to address particular administrative problems, chronological age evolved into a taken-for-granted element of personhood that we use as the basis for day-to-day interactions. Rather than arising naturally from everyday practice, chronological age became a privileged standard for classifying individuals because state bureaucracies formalized its widespread use as the standard to determine eligibility for public pension benefits. The adoption of chronological age as a metric for standards was not an end in itself. Its imposition served as a means to improve other processes; it brought numerical precision, certainty, and impartiality to classification practices that are otherwise inexact and arbitrary. Only later did the use and, indeed, knowledge of age in years become ubiquitous—achieving status as the classification employed in even grassroot standards. The schoolyard taunt "No five-year-olds can play" illustrates the use of a grassroots convention in everyday classificatory exercises of inclusion and exclusion.

Today, chronological age is remarkably wide in the scope of its coverage. On the basis of chronological age, women are referred for mammograms, children are allowed to play league soccer, and seniors qualify for dining discounts. In a host of contexts, from consumer marketing to curricular rubrics, chronological age serves as an operational measure of needs, capacities, preferences, and entitlements. The general rationale for these uses is the association of chronological age with other things that can be measured only imperfectly or with greater effort. Age in years, however, is so frequently and commonly employed that we lose sight of how loosely associated chronological age actually is with reading readiness, musical tastes, and other "age-related" factors.

Chronological age has supplanted other useful ways of thinking about age, a significant administrative achievement, particularly because most people in the West did not even know their exact chronological age before the eighteenth century. Greater precision in reckoning age is the result of embedding chronological age in administrative practice. When age really mattered in practice, individual precision in reporting chronological age swiftly followed. And, as chronological age became more significant to administrative processes, as it became the basis for standards, the production of exact age emerged as an objective in its own right. Although many things ride on chronological age, some people today pose challenges to age systems of classification. Either they do not know their precise age or their age records—collected over a lifetime for disparate purposes—are inconsistent. Because we invest so much in chronological age as the basis for a standard, bureaucracies find it necessary to develop standards for age, that is, formal rules to be followed in evaluating evidence that might serve as proof of age. In short, like Steven Epstein (chap. 2), Martha Lampland (chap. 5), and Martin Lengwiler (chap. 4 in this volume), I address here the broad classificatory project that enabled the standardization, both normative and statistical, of infinitely variable human beings.

Alternative Concepts of Age

When asked how old we are, we invariably answer with a number, our chronological age in years. We have other options, such as "younger than you are" or "older than I look," but these replies are generally taken as flippant retorts rather than responsive answers to a serious question. That we discount such responses suggests that chronological age has crowded out other legitimate ways of classifying age from our everyday consciousness. Although we may emphasize chronological age, this numerical indicator reflects only one conception of age. In many circumstances, there are other, more appropriate concepts than the mere passage of years. Thus, it is useful to consider alternative age concepts that have been overshadowed by our privileging of chronological age.

Relational, social, functional, and subjective criteria all speak to aspects of age that are imperfectly captured by the calendar time since birth. In traditional Confucian culture, where social relations were structured by hierarchical obligations, relational age—for example, being the older rather than the younger brother— mattered more than being thiry-nine instead of twenty-nine years of age. Older brothers merited the respect of younger ones, whatever their chronological age. In U.S. society, fine points of etiquette and protocol still revolve around relational issues of who is older and who is younger. When extended families gather for Thanksgiving dinner, coveted places at the dining room table are reserved for older kin while younger ones—sometimes into middle-age—are relegated to an improvised kid's table in the kitchen.

Of course, individuals may also be age-classified according to the age-related social roles that they assume. This classification embodies social age. At the beginning of the nineteenth century, British social workers fretted about teenage boys who labored as the main support of their parental family. In keeping with a boy's social age as the principal breadwinner, his grateful kin bestowed on him adult privileges not in keeping with his youth: kippers, tea, and the most comfortable chair in the house (Gillis 1974). In the west of Ireland in the 1920s, Conrad Arensberg and Solon Kimball (1968) observe that men in their forties were classified as "boys" if they were still single and under the authority of an aging father who had yet to turn over the farm to their management.

Functional age is concerned with the capacity to carry out those responsibilities associated with a particular phase in the life course. Social age addresses role performance; functional age is concerned with age-related capabilities. For example, girls are not regarded as women until they have, at a minimum, reached puberty. In deciding who should be classified as marriageable and who should not, what matters in most societies has been physical maturity, namely the biological readiness to conceive, carry a fetus to term, and give birth to a healthy child. To be sure, our laws fall back on chronological age to set a lower bound for marriage (Glendon 1989), but functional age remains an important, if often unacknowledged, force in social life. Whether we consider Pakistan (where marriages are arranged early) or the United States (where later love matches are the rule), women who reach menarche earlier also marry earlier (Udry and Cliquet 1982). Thus, functional age plays a role, even in societies where chronological age is the dominant standard for classifying individuals. Indeed, the formal measurement of functional age has received scientific attention (e.g., tests of reading readiness, biomarkers of "normal" aging), but these complex classification projects see application as standards only for particular groups (young and old) and in narrow domains (education and medical care) in which the measures are often benchmarked against chronological age norms. No doubt, functional age matters more to societies that do not rely on exact chronological age to sort individuals.

People do not necessarily identify with their chronological age. Bernard Baruch (1995) illustrated this fact of life famously when he declared "old age is always fifteen years older than I am." Subjectively, people may classify themselves as feeling younger or older than their years, consistent with indicators of their functional age (e.g., health), their social age (e.g., widowhood), and their relational age (e.g., having children) (Logan, Ward, and Spitze 1992). People build their lives consistent with their personal assessments of the age group to which they belong. When preteen girls began to identify with adolescents instead of children, for example, they dealt a financial blow to the Mattel Corporation by abandoning Barbie dolls for CDs and cosmetics. The loose coupling of subjective and chronological age is one sign of grassroots resistance to the hegemonic influence of numerical age within an age-stratification system that determines who gets which rewards (e.g., autonomy and respect).

There is an imperfect match of subjective and objective indicators of age (Logan, Ward, and Spitze 1992). This points to contradictions in a society that privileges chronological age over other conceptions of how old we are. That chronological age came to be imposed as the basis for *the* age standard reflects a modernist proclivity for both classification and standardization. Our reliance on chronological age also illustrates the tensions and challenges posed by systems that seek to rationalize personal biography.

Embracing Chronological Age

In earlier times, hardly anyone cared about a precise accounting of time since birth. Historically, few people in the West could reckon their ages. Only the birthdays of the nobility were regarded as important enough to warrant an observance. Because making a fuss about a mere birth date was so closely associated with royalty, Thomas Jefferson expressed his discomfort when citizens of the new republic took to celebrating the birthday of George Washington (Chudacoff 1989). Birthdays for the average citizen, the commoner, and the peasant went unrecorded and passed unmarked. Where there were baptismal records, they languished in the parish registry because exact age (as opposed to a rough estimate) was of little use to anyone. Given the high levels of infant mortality, a preoccupation with birthdays could only have been a painful reminder of loss for many parents. If there were annual markers for the passage of personal time, they were the seasons, the new year (Chudacoff 1989), or a personal saint's day.

China was not marked by age indifference (Thompson 1990). The Chinese calendar, based on the cycle of animals, made it easier to recall the year of one's birth. The widespread practice of astrology gave people a reason to care about when they were born. The Chinese did not routinely celebrate birthdays as we do in the West

today, but ages sixty, seventy, and eighty were important benchmarks commemo-
rated by the common people. Chinese age reckoning differed from that in the
West. At birth, an individual was assigned an age of one, and calculations were based
on a lunar year that was several days shorter than the solar year used in the West
(Shryock and Siegel 1973). Parenthetically, the Chinese reckoning of time on the
basis of the lunar year illustrates the nesting and articulation of the classification
and measurement systems behind standards. The history of how the Gregorian
calendar came to be widely adopted by Western societies (and, thus, available for
the computation of chronological age) is beyond the scope of this chapter.

Outside China, hardly anyone bothered reckoning chronological age. Most
people neither knew nor cared exactly how long they had lived. Commenting on
England, Pelling (1991, 81) concludes, "(W)hile there was a sharp awareness of
the different phases of life, including old age, and the legal definition of an idiot
was someone who could not tell his own age, there is little hope of people knowing their age *precisely* before the eighteenth century or even later." Illiteracy contributed to the general imprecision about years-since-birth, but this lack of
numerical age-consciousness also reflected the absence of community practices
that might have given salience to chronological age. "In general, arrival at a specific
age created no entitlement to charity, poor relief, poor law pension or medical
treatment" (Pelling and Smith 1991, 20). This is not to say that there were no instances when chronological age was codified. Although age provided no charitable
advantage, except perhaps the right of the old to glean, English enactments dating
from 1349 excused individuals at age sixty from military service, compulsory labor, and prosecution for vagrancy (Thane 2000). The concern, however, was with
functional age, that is, the capacity to fulfill these obligations. Because chronological age served as a rough indicator of capacity, it was enough that people were able
to offer some reasonable estimate of their age in years.

Just because people were inattentive to chronological age did not mean that
they were indifferent to age, particularly functional age. Broad age groups were
universally acknowledged. In "As You Like It," William Shakespeare (1600) enumerated seven stages of humans from "the infant, mewling and puking in the
nurse's arm," to the old man "sans teeth, sans eyes, sans taste, sans every thing." By
the Renaissance, a distinctive representational iconography emerged that por-
trayed life's stages as stair steps ascending to vigorous middle years and then de-
scending to decrepitude (Cole 1992). Despite the metaphorical power of age
imagery, in practice there were fewer meaningful age distinctions than in our own
society, and the divisions between age groups were less sharp.

Absent age-segregated social institutions such as retirement or schooling, young
and old shared the same social worlds (Ariès 1962). There was little reason to separate youths from adults until the advent of formal schools. Age-grading within
schools, grouping youngsters by chronological age in the classroom, did not
emerge as an educational practice in the United States until the nineteenth century

(Chudacoff 1989). Later, protective child-labor laws would prove the wedge that finally segregated children from the world of working adults. Historically, older people, too, were largely integrated into the broader community. Retirement did not exist as an institution. People worked as long as they were able to do so. To quit working was to become a burden on family or community and to suffer the stigma of dependence and decrepitude. Admitting paupers of all ages, the early poorhouse exemplified the age-integration of the broader society. Only after specialized institutions such as the orphanages for children, asylums for the insane, and out-relief for working-age adults siphoned off its younger population did the poorhouse turn into a de facto old-age home (Gratton 1986).

The historical absence of age-specialized social institutions underscores the fact that many discrete age groups, each with attributions of distinctive needs and capacities, are a classificatory project of modernism. To be sure, age-graded societies have long existed, but social age defined by membership in a promotion cohort, not chronological age, was the hallmark of these societies (Levy 1996). Ariès (1962), the French historian, observed that adolescence blurred with childhood until the eighteenth century. The transitional age between childhood and adulthood was not fully institutionalized until 1904 when G. Stanley Hall, a U.S. psychologist, introduced the term adolescence in his landmark study *Adolescence: Its Psychology and Its Relations to Physiology, Anthropology, Sociology, Sex, Crime, Religion, and Education.* Since that simpler time, *preteen* has entered our vocabulary to mark the transitional phase inserted between childhood and adolescence. At the other end of the life course, gerontologists first distinguished between the distinctive needs and capacities of the "young-old" (65–74) and the "old-old" (75 and older) (Neugarten 1974). With the rapid aging of the older population, scholarly attention at the end of the twentieth century was soon directed to the "oldest old" (Bould, Sanborn, and Reif 1989) and to centenarians (Krach and Velkoff 1999). Despite heroic efforts to categorize the years between young adulthood and old age in terms of developmental stages, no particular categorization has taken hold, and middle age remains ripe for further compartmentalization.

Chronological Age and Old Age

Many factors contributed to the increased awareness of precise chronological age. Noting wryly that medieval Catholics were preoccupied with eternity, not longevity, Cole (1992) credits the modernist penchant for self-improvement with drawing attention to calendar age. Advanced chronological age, he argues, served as a rational indicator of a life lived in accord with moral principles and hygienic practices. Kohli (1986) offers another explanation for why broad age categories were supplanted by chronological age, a more precise measure that served to standardize the life course by facilitating the spread of age-specific rules, norms, and expectations. Critical to his argument is the emergence of the individual—liberated from family, church, and community—as the object for a host of state social programs, particularly those organized around the labor system. These arguments about the significance of work and health to the salience of chronological age direct our attention to old age. Until fairly recently, functional age and exact chronological age were only loosely coupled in the last phase of the life course, but the link between advancing age and infirmity has been tightened.

To limit demands on public funds and to encourage private responsibility, public assistance for the old was historically restricted to those who qualified by virtue of need (e.g., disability, inability to support self, and lack of kin) and merit (long time community residence, moral character, or faithful service). An arbitrary criterion based on chronological age might be used to establish eligibility, but this could only have been a rough benchmark of superannuation, given that most people knew only their approximate chronological age. Whether an eligibility criterion was set at sixty or eighty, it served to communicate generally shared understandings about the level of infirmity that might justify public support. In other words, the chronological age criteria that were applied in the administration of public old-age support spoke to functional age, the degree of physical or mental incapacity regarded as typical among those attaining a given chronological age. Although these judgments sometimes incorporated expert opinion and scientific understandings, they often built on lay beliefs (e.g., about the threshold of old age) that arose from everyday interactions of individuals within diverse social contexts.

The ready substitution of chronological and functional age criteria presumes the existence or construction of norms, be they social or statistical, about the capacities of individuals of different numerical ages. Shared understandings about the physiological declines associated with old age had, no doubt, always existed, but David Hackett Fischer (1978) notes the popular devaluation of older Americans that began in the earliest days of the Republic. Negative cultural views of old age were reinforced by a growing body of scientific knowledge that lent empirical specificity and statistical precision to age norms. Based on scientific observation, physicians in Parisian hospitals were among the first to draw on clinical experience and autopsies of elderly patients to document the progressive physical degeneration associated with the passage of years (Troyansky 1989). By the mid- to late nineteenth century, British and U.S. doctors were also describing old age as a pathological condition marked by unique organic deterioration (Haber 1983).

The belief in age-related decline justified the creation of broad categories of incapacity bounded by chronological age, as suggested by the physician, William Osler, in 1905. Upon his retirement from Johns Hopkins University Medical School at the age of fifty-eight, Osler delivered an influential valedictory that lauded the productivity and creativity of men ages twenty-five to forty, regretted the comparative ineffectiveness of men over forty, and decried the complete uselessness of those over sixty years of age (Cole 1992). U.S. medicine's pessimism regarding the capacities of the aged gave fuel to a broader social project that resulted in the denigration of the old and their eventual withdrawal from the labor market. Physicians, however, were hardly alone in the scientific study of age-related differences in (in)capacity. Among those who documented statistical age differences in performance were Adolphe Quetelet, a Belgian mathematician; Sir Francis Galton, a British scientist, and G. Stanley Hall (Chudacoff 1989). Thus, the fruits of scientific rationality offered a justification for the administrative use of chronological age to create a standard for attributing incapacity to older adults.

U.S. Civil War Pensions: Commensurating Age and Incapacity

Public old-age support demonstrates the historical emergence of time-since-birth as the standard measure of age subsuming all other concepts. Lampland (chap. 5 in this volume) points to the way in which administrative practice standardized Hungarian laborers for work. Tellingly, administrative practices have also embraced chronological age as a process to standardize individuals efficiently and equitably for a morally legitimated withdrawal from work. In creating age-stratified access to rights, responsibilities, and rewards, public systems of old-age support are critical to institutionalizing chronological age. Perhaps no institutional development embodies the creeping equation of chronological age and physical incapacity as clearly as the U.S. system of Civil War pensions (Treas 1986). In 1862, Congress passed legislation to provide for claims that arose from deaths and disabilities occasioned as a direct consequence of military service for the Union (Glasson 1918). Pension benefits were standardized—matched to the seriousness of the disability, as specified by lists of qualifying wounds and diseases. Although eligibility initially depended on objective criteria such as the loss of a limb or total blindness, eligibility was gradually broadened to include more ambiguous functional criteria that reflected general health conditions. For example, $15 per month was to be paid for a disability equivalent to the loss of a hand or foot, $20 for the inability to carry out light manual labor, and $25 for a disability that required the regular assistance of another person. The normal, sighted, healthy male with four intact limbs (from which pension beneficiaries deviated) stands as a forbearer of the standard human whose elaboration Epstein (chap. 2 in this volume) analyzes.

In the case of chronic illness, local boards of examining physicians drew on their expertise to rate each disability with fractional exactitude. Incapacity was calibrated in terms of eighteenths of the disability for manual labor equivalent to the loss of a hand or a foot. This precision was illusory, serving largely to obscure the lack of uniformity among local experts. According to the foremost historian of these pensions, William H. Glasson, the difficulties of the task and the variable expertise of the examiners were apt to result in "either the arbitrary or the absurd—and sometimes both" (1918, 138). One claimant who was rated independently by four local medical boards was judged to have no ratable disability by one board but was rated for disabilities worth $8, $17, and $24 per month by the other three (White 1958, 211). In short, the efforts to standardize disability ratings floundered on the inability to standardize the raters' practices.

Because veterans were an important political constituency, politicians were sympathetic to demands to liberalize eligibility and expand benefits. In 1890, legislation extended eligibility to anyone who had served in the U.S. army or navy for ninety or more days during the Civil War, who had been honorably discharged, and who suffered from a "mental or physical disability of a permanent character, not the result of their own vicious habits, which incapacitates them from the performance of manual labor in such a degree as to render them unable to earn a support" (Glasson 1918, 234). The new law guaranteed a pension to all veterans, except those who had health that was "practically perfect" (Sanders 1980, 142). Thus, Civil War pensions became a system of old-age support.

The commissioner of pensions was authorized to set benefit rates for disabilities that were not directly addressed by the law (Costa 1998). By 1873, the list of rated disabilities served not only to specify a dollar benefit but also as a comparative standard to determine the worth of disabilities not otherwise classified. Although the 1890 legislation did not mention age, the Pension Bureau applied age-based eligibility criteria. The commensuration of chronological age and functional age was clear in the instructions issued to the examining physicians by the commissioner of pensions: "A claimant who has reached the age of 75 years is allowed the maximum rate for senility alone, even where they are no pensionable disabilities" (Glasson 1918, 243). Even younger applicants could be readily determined to be disabled. "A claimant who has attained the age of 65 years is allowed at least the minimum rate, unless he appears to have unusual vigor and ability for the performance of manual labor in one of that age" (Glasson 1918, 243). Emphasizing that old age was but another medical condition, the instructions stated that "(t)he effect of partial senility should be considered with other infirmities, where there are such, and the aggregate incapacity stated" (Glasson 1918, 243).

Thus, the administrative practices of the Pension Bureau equated *chronological* age with a specified degree of disability that, in combination with minor disabilities, would qualify the older veteran for a pension. By 1904, President Theodore Roosevelt's Executive Order #78 provided that a simple declaration of old age was sufficient evidence of incapacity, even in the absence of other disabilities. The Service and Age Pension Act of 1907 gave congressional blessing to this practice. Advancing chronological age was equated with increasing degrees of infirmity so that veterans sixty-two to sixty-nine were paid $12 monthly, those seventy to seventy-four were paid $15, and those seventy-five and older were paid $20.

The reach of pension eligibility and benefit standards based on chronological age classifications cannot be overstated. According to estimates by social reformer Isaac M. Rubinow (1913), two-thirds of native-born, white, 65-year-old men living outside the South were drawing a federal veteran's pension in 1913. In 1909, the average annual Civil War pension amounted to 29 percent of the average annual earnings of factory workers in the United States (Treas 1986). Beyond sensitizing hundreds of thousands of pension recipients to the importance of chronological age, the use of chronological age served to legitimate rewards for one class of citizenry, even as it excluded Confederate veterans, former slaves, and recent immigrants. The Civil War pensions enabled aging veterans to withdraw from the labor force and to live independently of kin, thus marking the beginning of the modern institution of retirement (Costa 1998).

In embracing chronological age as the standard measure of late-life infirmity for Civil War veterans, the United States was in step with Germany, Great Britain, Denmark, and other states that imposed chronological age on their national old-age pension systems. Although most European states set a higher threshold of old age than did the U.S. Civil War pensions (Skocpol 1992), the cross-nation comparison is apt. Scholars point to the Civil War pension to explain why the United States lagged three decades behind Great Britain in establishing a general system of old-age support. The veteran's pensions deflected political pressure for a universal system while fanning Progressive Party resistance to the political patronage and corruption it bred (Ikenberry and Skocpol 1987; Orloff 1988). Observing that the U.S. veteran's pensions cost three times more than the needs-based old-age pension of Great Britain, Rubinow (1913, 405) charges that the Civil War pensions constituted a covert social insurance scheme, "an economic measure which aims to solve the problems of dependent aged and widowhood."

More significantly from the perspective of standards, Union army pensions legitimated the administrative use of chronological age as an item of personal biography by embedding it in the everyday practices of an enormous and influential bureaucracy. Said by its commissioner in 1891 to be the world's biggest executive bureau, the Pension Bureau employed 3,000 clerks in the nation's capital, 3,800 examining physicians throughout the United States, and 400 agents in eighteen regional offices. Although U.S. Civil War pensions were not a universal entitlement, their remarkable reach and range magnified the importance of an administrative expedient—the impartial use of years of age to gauge incapacity, infer need, and establish entitlements. Chronological age was, therefore, the basis of an imposed standard promoting exactitude and regularizing the haphazard business of determining pension eligibility. The enormous scope and scale of chronological age as an administrative practice guaranteed that years in age had personal implications for millions of Americans.

Old-Age Pensions in Ireland: Making Age Matter

The year after the U.S. Congress wrote chronological age into Civil War pension legislation, the British Parliament established an old-age pension for British subjects ages seventy and older residing in the United Kingdom. The U.S. system demonstrates how chronological age, functional age, and benefit amounts were coordinated and commensurated in the context of an enormous bureaucracy and idiosyncratic local practices. The 1908 Old Age Pension Act in Britain demonstrates the impact that eligibility standards based on chronological age had on age-consciousness. In Ireland, a country characterized by extreme poverty, the pensions were coveted. According to one farmer, the pensions meant not only additional income but also added years of life. "To have old people in the house is a great blessing in these times because if you have one, it means ten bob a week and, if you have two, it means a pound a week coming into the house. . . . Not only that but it adds at least ten years to a man's life because the anticipation that each Friday he is to get ten shillings will cheer him up and keep him keen" (quoted in Arensberg and Kimball 1968, 120). Despite the strong incentive to qualify for a pension, establishing age-based eligibility proved difficult. Ireland did not have a civil birth registry prior to 1864. In the absence of proof of birth date, there was invariably a temptation to add years to life in order to qualify for a benefit.

Although church records were acceptable ways of establishing age, few people had these documents. Administrators were called on to determine age on the basis of such dubious criteria as physical appearance, local testimonials, or the individual's ability to recollect some significant historical event or natural disaster, such as a memorable "great wind." Comparing the censuses of 1901 and 1911, John Budd and Timothy Guinnane (1991) brilliantly demonstrate that the 1908 legislation was followed by increased awareness of chronological age. People who were once happy to round their ages to the nearest "0" or "5" moved toward greater accuracy in their age reporting. This greater precision characterized those we know to have been most vulnerable to age-heaping and casual reporting, namely, the old, the illiterate, and the laboring poor. Whatever they lacked in the way of personal resources to facilitate knowledge of chronological age, these people—being the most likely to qualify for a means-tested, old-age pension—were strongly motivated to employ an exact numerical age when they interacted with the welfare state. State interventions attaching importance to chronological age—the basis for an imposed standard—achieved a revolutionary change in the everyday accounting of personal biographies. Undoubtedly, they can also be credited for the standardization of the life course to the extent that workers began to time their labor force withdrawals to coincide with pension eligibility schedules and emergent retirement norms.

Age Exactitude

As the Irish case demonstrates, the use of chronological age for administrative purposes presupposes an administrative apparatus or some customary practice for recording all birth dates. Until the twentieth century, many births went unrecorded in the United States. Despite local record-keeping and informal entries in the family Bible, there was no national system or universal custom to assure that every birth date would be known so that every American could calculate an exact chronological age. With age in years embedded in administrative practice, the creation of chronological age as a reliable product became increasingly important. Because recording vital events was the business of the states, the federal government was late to standardize data collection practices by setting uniform criteria, such as a standard birth certificate. When the U.S. Bureau of the Census established the Birth Registration Area in 1910, only ten states and the District of Columbia registered a sufficiently high proportion of births to qualify for membership (Shryock and Siegel 1973). By 1933, all states were part of the registry, reporting births to the federal government, but some births, such as home births to black women in the rural South, still sometimes went unreported. In light of the delays in establishing and fully implementing the administrative mechanisms to establish chronological age, it is not surprising that even today some U.S.-born Americans are not entirely certain about their birth date and, thus, their exact age. In the past, the inability to report an exact age was much more widespread and came to be recognized as a serious problem for the statistical infrastructure of the United States.

Although the United States did not move to ensure uniform recording of births until the twentieth century, the censuses of population permit us to trace the federal government's increasing desire for age exactitude. Changing Census Bureau protocols reveal the growth of bureaucratic concern and scientific expertise regarding the citizens who could not or would not give their exact age in years. Federal interest in chronological age dates back to the earliest census of 1790. As table 3.1 demonstrates, census-takers over the decades sought to achieve greater detail and exactitude in age reporting, even though many of those enumerated could do no more than guess their chronological age. Only the most privileged segment of the population—free white males—warranted even a crude age breakdown in 1790. In keeping with the growing importance of chronological age, successive censuses extended the collection of age data to all classes in the population, including women, slaves, free blacks, and Indians.

In the first census of 1790, free white males were enumerated as being younger than sixteen and as being sixteen or older (Gauthier 2002). By 1830, when the first uniform printed schedules were employed to collect data, free white people—men and women—were tabulated in thirteen age categories ranging from younger than five years to one hundred years and older. The ages of slaves and free blacks were also tallied in six categories ranging from younger than ten years to one hundred years and older. The choice of categories reveals an official lack of sophistication about the age distribution of the U.S. population and the likely inaccuracy of age reports. The 1830 census schedule sorted out free white octogenarians, nonagenarians, and centenarians, respectively, signaling a strong taste for precision. Given the high mortality of the era, the very old, however, were rare in the popu-

Table 3.1. U.S. Census data collection on chronological age, 1790–2000

Census year	Age data collected	Enumerator instructions
1790	16 and upward, under 16	Free white males only
1800	Under 10 years of age, 10 and under 16, 16 and under 26, 26 and under 45, 45 and upward	Free white persons only
1810	Same as 1800	
1820	Same as 1800	Free white persons
	Plus, 16 and under 18	Free white males only
	Under 14, 14 and under 26, 26 and under 45, 45 and upward	Free colored persons and slaves
1830 (First uniform schedule)	Under 5, 5–10, 10–15, 15–20, 20–30, 30–40, 40–50, 50–60, 60–70, 70–80, 80–90, 90–100, 100 and upward	Free white persons
	Under 10, 10–24, 24–36, 36–55, 55–100, 100 and upward	Slaves and free colored persons
1840	Under 14, 14 and under 25, 25 and upward	Deaf and dumb white, slaves, and colored persons
	Same as 1830	
1850	Specific age of each person at his or her last birthday. Fractional parts of year for child under 1 year (one month, one-twelfth).	"If the exact age in years cannot be ascertained, insert a number which shall be the nearest approximation to it."
1860	Same as 1850	"Where the age is a matter of considerable doubt, the assistant marshall may make a note to that effect."
1870	Same as 1860	
1880	Same as 1870	
1890	Age at nearest birthday in whole years; for children less than 1 year on June 1, age is in twelfths of a year.	"Do not accept the answer 'Don't know,' but ascertain as nearly as possible the exact age of each person. The

(continued)

Table 3.1. (continued)

Census year	Age data collected	Enumerator instructions
		general tendency of persons in giving their ages is to use round numbers, as 20, 25, 30, 35, 40, etc. If the age is given as 'about 25,' determine, if possible, whether the age should be entered as 24, 25, or 26."
1900	Date of birth	"The object of this question is to help in getting the exact age in years of each person enumerated. Many a person who can tell the month and year of his birth will be careless or forgetful in stating the years of his age, and so an error will creep into the census. This danger cannot be entirely avoided, but asking the question in two forms will prevent it in many cases."
	Age at last birthday in completed years. Age in completed months for child under one year of age.	"An answer given in round numbers, such as 'about 30,' 'about 45,' is likely to be wrong. In such cases endeavor to get the exact age."
1910	Age in completed years at last birthday prior to April 15. For child less than two, age in twelfths (e.g., 1 3/12).	"(W)hen an age ending in 0 or 5 is reported, your should ascertain whether it is the exact age. If, however, it is impossible to get the exact age, enter the approximate age rather than return the age as unknown. . . . Some difficulty may be met in ascertaining the exact ages of Indians, as they frequently reckon their ages from notable events occurring the history of the respective tribes. Endeavor to ascertain the years in which these notable events occurred, and with a little calculation on your part you should be able to ascertain the exact age of each Indian."
1920	Same as 1910, except age for child under 5 to be reported in complete years and months.	

1930	Same as 1920
1940	Same as 1930
1950	Same as 1940; if under one, birth month entered
1960	Month and year of birth; age at last birthday
1970	Same as 1960
1980	Same as 1970
1990	Age and year of birth
2000	Age on April 1, 2000; month, day, and year of birth

Only quarter of birth transcribed

Source: Jason G. Gauthier. 2002. *Measuring America: The Decennial Censuses from 1790 to 2000.* Washington, D.C.: U.S. Census Bureau.

lation. Many, if not most, individuals falling in these categories of advanced old age were undoubtedly old people vastly overstating their years, unbridled by birth records or by strong conventions demanding that a person know and report his or her age exactly.

In 1850, the census of population moved away from age categories to require individuals to report numerical age as of the last birthday. In the 1890 census, the age item—asking for the age at the nearest, not last, birthday—proved confusing and was subsequently abandoned. In a further quest to achieve precision in age reports, specifically to prevent "careless or forgetful" reports of age, the 1900 census inquired not only about age at last birthday but also about date of birth. After abandoning the query about birth date in 1910, the Census Bureau in 1960 returned to asking about birth date (i.e., month and year) as well as age. In 1990, the census asked only the year of birth in addition to age. Census 2000, however, extended the quest for greater and greater precision in chronological age into the twenty-first century. In addition to determining age on April 1, 2000, the 2000 census asked for month, day, and year of birth.

Recognizing that some people did not know their exact age, government officials developed standard practices to deal with the problem. In 1890, enumerators were encouraged to enter an approximation. The census-takers' instructions cautioned that people round their ages to the nearest 0 or 5, and the enumerators were urged to reject "about 25" in favor of a more accurate answer. In 1910, there were special instructions for Indians, who sometimes gauged their ages with reference to historical events. Anticipating the "historical calendar" adopted by later demographers working in less-developed countries (Scott and Sabagh 1970), enumerators were told to determine the year of any tribal event mentioned in order to calculate an exact age.

Acknowledging the errors in age reports, the Bureau of the Census published its first comprehensive study of age statistics in 1904 (Young 1904). As the study pointed out, accurately classifying the population by age was essential to the work of the state. In addition to gauging the numbers of potential voters, workers, and soldiers, these statistics provided estimates of fertility (in the absence of systematic birth records) and contributed to the assessment of poverty, crime, literacy, and mortality. Serious analyses of errors in age reports by sex, region, race, and nativity were presented. So were population tabulations by single years of age—carefully adjusted to smooth out the irregularities introduced by rounding, heaping, and other reporting errors. The publication not only made clear the importance of accurate age reports but also demonstrated for the first time how demographic expertise in scientific observation and statistical analyses could be harnessed to improve the quality of census age data as it was solicited in the field and tabulated in the office.

While chronological age was becoming the basis for the standards for the distribution of rights, responsibilities, and entitlements, the state was also at work to create the birth-reporting infrastructure that would permit future generations to establish exact age. The Census Bureau, however, was charged with collecting age data even before literacy, the birth-registration system, and the public's age-consciousness were sufficiently widespread to assure that all people knew their exact chronological age. Successive censuses demonstrated the growing importance of chronological age by extending age inquiries to the entire population and by demanding progressively more detailed information. They also showed an increasing sensitivity to the seriousness of errors in age reports. Census-takers received formal instructions on how to assign ages to those who did not know their age and how to encourage greater accuracy from those who did. Accuracy remains a concern with the census today, and considerable effort is devoted to formatting the self-administered questionnaires so as to avoid careless errors and misunderstandings. Although contemporary Americans still make errors in reporting age, chronological age is now a standard product and uncertainty about one's chronological age is relatively uncommon.

Breakdowns

It does not really matter whether people know their chronological age unless they bump up against bureaucratic systems that demand chronological age. In telling the story of refugees in California, Anne Fadiman (1997) contrasts a Hmong woman's limited knowledge about her age with her daughter's birth in a U.S. hospital—a birth that is recorded meticulously and to the minute. Although this medical quest for age exactitude extends back to the time of conception, it was not possible to assign a gestational age to the newborn because the mother had not received any prenatal care. When she delivered her baby, the mother's birth date was recorded as October 6, 1944. On other hospital visits, her birth date was recorded as October 6, 1942, or October 6, 1926. Unclear as she was about the year, this Hmong mother was very certain about the month. Marking events by agricultural periodicities, her parents had informed her that she had been born in the season when farmers stacked their rice stalks and gave their opium crops a second weeding. Certainly, age would have been a salient concern in her village, but functional age (e.g., being old enough to help with crops), not chronological age, was what mattered. The mother's birth year was a fabrication, a matter of indifference in her Laotian village, but a necessity in U.S. society, where everyone must know his or her birthday.

The Hmong case serves to demonstrate that the requirements for age exactitude do not always demand age accuracy. The hospital staff apparently failed to notice or care about the inconsistencies in the birth dates they had recorded. Nobody noted that a 1926 birth year implied that the mother was fifty-five years old at the birth of her next-to-last child. Nor, it seems, did any U.S. immigration official question the Hmong refugee who reported July 19 as the birth date for each of his nine children, albeit for nine consecutive years (Fadiman 1997). In everyday practice, even the best practitioners, charged with collecting the raw material of age classification, must work around the occasional uncertainty or inconsistency. Any vaguely reasonable, if highly specific, date is sufficient to meet most bureaucratic requirements. The expedient work practice is to ignore all but the most egregious inconsistencies in ages or birthdays. This practice is facilitated by the fact that few people are on the look out for fabricated ages, given the widespread conviction that everyone knows his or her chronological age.

The Hmong people employed the uncertainty about their ages to strategic ends. In Thai refugee camps, they understated the chronological age of elderly parents under the (mis)belief that the United States would turn away older immigrants (Fadiman 1997). They overstated the ages of their children in the refugee camps so that they could get larger food rations, but in the United States, they understated the youngsters' ages so that they could remain in school longer. Who is there to object when age is truly unknowable or when the costs of obtaining precise information are much higher than what might be gained from the knowledge? Sometimes, however, the stakes are higher, the problems more frequent, and informal work practices will not suffice. Sometimes, more formal patches must be crafted for the classification systems founded on chronological age standards. When age is the basis for important standards, it is necessary to have standards for age.

Standards for Age

What happens in an age-conscious society when an individual does not know his or her exact chronological age—or at least cannot establish it to the satisfaction of authorities? Given the centrality of chronological age as a classifying mechanism underpinning standards, particularly in administrative contexts, there must be standard operating procedures to deal with cases in which ages are unknown, unreported, inconsistent, or implausible. The format and wording of items on questionnaires and administrative forms are subject to careful evaluation to increase response rates and improve the reliability of age reports. National statistical bureaus employ various methods to smooth the lumpy aggregate age distributions that arise when census respondents round their ages to culturally preferred digits (Shryock and Siegel 1973). In statistical analyses, individuals who fail to report an age will have an age imputed or assigned to them, according to an algorithm based on the average age of the population, the age of people of like characteristics, the ages of related household members, and so on. In some instances, the age of individuals must be known with a high degree of certainty, and it is not sufficient to rely on self-reported age alone.

As proof of age for benefit eligibility, the U.S. Social Security Administration requires evidence of a person's birth date. The preferred evidence is that which is recorded prior to the fifth birthday in the form of either a public birth record or a religious record of birth or baptism. Absent this preferred documentation, Social Security lists eighteen other documents that may be used to establish date of birth. As shown by the following list, the diversity of records giving birth date or age is quite extraordinary. Collected for a host of educational, commercial, private, medical, religious, civil, military, and economic purposes, these documents underscore the widespread use of exact chronological age and birth dates in U.S. life. Ironically, the Census Bureau, the agency that long struggled with its own problems in determining exact age, issues official transcripts from its earlier census records that may be used to establish age in order to qualify for social security and other retirement benefits. Social Security's list of potential age evidence is not exhaustive, and other detritus of the life course may also be used to validate the essential biographical information of chronological age.

Records that may be submitted to the Social Security Administration to prove date of birth (Social Security Administration 2001)

- Public record of birth established before fifth birthday
- Religious record of birth or batism established before fifth birthday
- School record
- Census record
- Bible or other family record
- Religious record of confirmation or baptism in youth or early adult life
- Insurance policy
- Marriage record
- Employment record
- Labor union record
- Fraternal organization record
- Military record
- Voting record
- Vaccination record
- Delayed birth record
- Birth certificate of child showing age of parent
- Physician's or midwife's record of birth
- Passport
- Immigration record
- Naturalization record

The layering of evidence on age—collected over the life course by different parties for disparate purposes—introduces the possibility of conflicts and inconsistencies between information from different sources. Because it is necessary for people to establish chronological age in order to establish benefit eligibility, the Social Security Administration has developed administrative mechanisms as well as evidence policies. These are designed to adjudicate the breakdowns in classificatory procedures that arise from uncertainties or conflicting evidence about the chronological age of a claimant. For example, a long-standing practice has been to give the greatest weight to the oldest records of chronological age. Whatever their original purpose, the earliest layers of age records are least likely to be tainted by the expediency of meeting the eligibility requirements for old-age pensions. (As we will see, age may certainly be falsified early in the life course to meet other objectives.) The probative value of different records has increasingly been grounded in technical expertise. Since 1968, the relative validity and reliability of various types of documents has been determined with statistical research based on Evaluation and Measurement System (EMS) data. Despite this rationalization of practices, procedures, and policies, some cases are not resolved by mechanical decision-making rules.

Administrative case law offers a window on the landscape of chronological age as lived by individuals, recorded by sundry agencies, and adjudicated by the courts. In the case of "W," recent evidence of chronological age was favored over the earlier records normally preferred (Social Security Administration 1960). Indeed, relational age was used to establish a birth date that the claimant had hardly ever asserted over her long lifetime. In 1957, "W" sought to prove that she was born in 1895 and had reached sixty-two, the minimum eligibility age for a widow's pension. Over the years, she had claimed an 1899 birth date, notably in her earlier Social Security application for benefits as the widowed mother of a dependent child. Immigration arrival records from 1911 listed her as eleven years of age, implying that 1899 was indeed her birth year. However, "W" explained that her mother had passed each of her six children off as younger in order to save money on their steamship fares. Backing her contention was the fact that her youngest sibling must have been older than the age of three recorded at arrival since her mother and father had been separated for five years. Affidavits from her siblings backed up her claim that she was born in 1895. Although "W" was the second-born child (her birth order was established in the 1911 arrival records), birth dates from the school records of the fourth, fifth, and sixth children ranged from 1899 to 1906, suggesting that "W" had been born sometime before 1899. Last, her optometrist certified that "W" had owned up to a birth date of December 24, 1895, but only because he had insisted "that she tell him her exact age because of the medical consequences" (quoted in Social Security Administration 1960, 77). Thus, in a life course in which birth date was apparently marked by both strategic use and selective salience, a careful parsing of information on chronological and relational ages led to a rejection of the bureaucratically preferred early records in favor of more recent evidence.

In addition to inconsistencies over time, individuals may have more than one officially recognized chronological age depending on the jurisdiction. One claimant for old-age benefits produced two decrees from Greek courts establishing his birth date as June 10, 1893 (Social Security Administration 1965). The Social Security examiner determined the claimant's birth date was June 10, 1901, based on the biographical information he had given a number of times over three decades. Believing that the Greek proceedings were undertaken to bolster the claimant's application for benefits, the U.S. District Court held that the Social Security Administration, although obligated to weigh the decrees as evidence, was not bound by Greek court proceedings to which it had not been a party.

Whereas some age uncertainties are handled on a case-by-case basis, others constitute a large enough problem to require a formal bureaucratic patch for an entire class of individuals. For example, the Social Security Administration has had a special policy for determining the age of Holocaust survivors (Social Security Administration 1981). During World War II, the Nazis sentenced persons, ages eighteen to thirty, to work camps and others to death camps. Some victims of Nazi persecution misstated their ages in order to survive. Later, they continued to use these incorrect birthdates, even on official records of their immigration to the United States. When birth records were lost or destroyed in the war, these immigration records constituted the earliest (and generally preferred) evidence of age. Recognizing the hardship posed for the elderly who could not demonstrate their eligibility, the Social Security Administration implemented a special procedure. An individual first established his or her status as a Holocaust survivor (e.g., with such evidence as tattoo numbers, German passports stamped with a red J, or evidence of wartime residence in one of twenty-four countries as a member of a group marked for extermination). If the Social Security Administration was unable to obtain preferred evidence of the Holocaust survivor's age from the country of birth, then statements of friends, family, or even the claimant were deemed acceptable to establish true age.

In short, unique biographical and historical situations render some individuals "ageless." These uncertainties about chronological age constitute breakdowns for classification systems that rely on chronological age as the basis for standards that assign rights, verify entitlements, and enforce responsibilities. Although an approximation of exact age may be adequate in some instances, there are other circumstances in which ambiguity must be eliminated, even at some cost to the participants. As in the case of the Social Security Administration, when birth records are unavailable, a host of other records may come into play to establish an official, and hence "true," chronological age. The problem of assigning age, however, is by no means limited to pension systems, older adults, developed countries, or written documents for adjudication. In the less-developed world, for instance, the law may specify a medical examination to determine whether the age of a young victim is consistent with a criminal charge of child sexual abuse (where "child" is based on some chronological definition such as "less than sixteen years") (Republique Democratique du Congo 1970). In a host of circumstances around the globe in which the knowledge of chronological age is essential, the elusiveness of exact chronological age elicits procedural patches that facilitate the functioning of age-based systems. Chronological age as a practice demands attention to chronological age as a product.

The Triumph of Chronological Age

Once a fact known only by the privileged few, an individual's chronological age emerged as an essential piece of information standardizing the personal biography. Today, chronological age determines the timing and progression of individual lives via the many informal age norms and formal age rules that link people to age-graded social institutions. In the past, these social norms were based on shared understandings of functional and relational age differences, but chronological age has come to be preferred over other measures of age-based (in)capacities, needs, and resources. Chronological age has been naturalized and is now so widely accepted as a rightful delimiter that it is invisible even as it is omnipresent. It is scarcely contested. Consistent with the moral power of standards noted by Lengwiler (chap. 4) and Lampland (chap. 5 in this volume), the use of chronological age in various standards embodies moral assessments and brings moral legitimacy to a variety of practices.

Although age systematically organizes the relationships of individuals within a host of social institutions across the life course, the historical movement to greater age exactitude in the United States must be read as the triumph of the imposed age-based standards traced to the state and administrative practices. Embedded in administrative processes, the state also implemented the universal system for birth recording that provides Americans with an essential product—incontrovertible evidence of their date of birth. The processes of the state, however, established the incentives to reckon age accurately.

Although my discussion has emphasized chronological age as a classificatory criteria establishing the standards for pension eligibility, exact age is also used to classify individuals into those who can and cannot vote, drive, and be tried as an adult in criminal court. It identifies people who are or are not obligated to register for the draft or to attend school. As testimony to the grassroots adoption of the years-of-age standard, chronological age in personal ads proclaims who is and who is not an acceptable romantic partner. By conferring rights and responsibilities, chronological age constitutes a stratification regime, albeit one in which mobility is systematic and predictable—assured by the passage of years. In concert with social institutions, chronological age standardizes the course of American lives. This is hardly a unique phenomenon but, rather, an instance of a more general process explored throughout this book; contrary to an individualizing trend evident since the Enlightenment, human beings are routinely classified and standardized—by chronological age as surely as they are for labor (Lampland, chap. 5), for insurance eligibility (Lengwiler, chap. 6), and for medicine (Epstein, chap. 2 in this volume).

Remarkable in its scale and scope, chronological age can standardize personal biographies because age is the metric used in standards in diverse contexts. Proof of age, for instance, is essential to establishing eligibility for Social Security and other old-age benefits. Because age is the basis for standards incorporated in rationalized rules of decision making, having a chronological age that is unknown or inconsistent causes a breakdown of classificatory mechanisms. Despite birth registries that systematize age information, breakdowns occur, either because some people (e.g., refugees) have irregular personal biographies or because the use of chronological age in multiple contexts results in a layering of contradictory evidence. If age is the standard, there must be standards for age, that is, regularized procedures to deal with uncertainties. This exemplifies the networked character of standards noted by Susan Leigh Star and Lampland (chap. 1 in this volume). Basing benefit eligibility standards on chronological age, the Social Security Administration not only issues a long list of documents that may be used to prove date of birth but also maintains evidentiary policies and procedures to adjudicate among conflicting records. Even with this degree of bureaucratic rationality, there are cases that can only be resolved in an administrative or court hearing. Thus, exact chronological age—determined on the basis of formal probative standards—becomes the classificatory linchpin of a host of standards for stratifying individuals in a uniform fashion—a practice that gives rise to the standardization of the life course.

■ Coffins Expand with Occupants

YORK DAILY RECORD/SUNDAY NEWS

Charleston Daily Mail (West Virginia)   *November 1, 2006*
Section: News

America's obesity driving up size of final resting places, morgues
York, Pa.—Coping with America's growing obesity doesn't end with death.

York County Coroner Barry Bloss recently said York Hospital's laboratory is considering renovating its morgue because bodies of more than 300 pounds can't fit into the drawers.

Because of the increasing number of obese bodies, the hospital is considering switching from its drawer system to a refrigerated room where the deceased can be wheeled in on litters.

Apparently, it's a growing problem for those dealing with death.

"There really is not an area untouched by this trend," said Scott McSparran, the marketing director for The York Group, a local maker of coffins.

To make room for our expanding waistlines, the wooden-casket manufacturer will increase the interior width of its 2007 standard-sized model from 22 inches to 24 inches.

Unlike, say, airplane seats, the expansion isn't necessary for user comfort.

"But we don't want to give the appearance of the deceased being crammed into a casket," McSparran said.

The Centers for Disease Control and Prevention has termed America's growing obesity as "an epidemic." According to data from the National Center for Health Statistics, 30 percent of U.S. adults 20 years of age and older—more than 60 million people—are obese, or more than 30 percent over their ideal weight.

Those statistics now follow us to the grave.

The York Group is the second largest manufacturer of caskets in America, and it is a leading manufacturer of all-wood caskets. It's owned by Pittsburgh-based Matthews International Corp., a leading manufacturer of bronze memorials.

Since 2002, sales of wood oversize caskets have tripled—a sale trend McSparran expects will continue to grow.

The standard size, he said, "just doesn't do it for a lot of our customers anymore."

In 1930, the interior width of the company's typical coffin was about 21 inches, McSparran said.

Today, Matthews makes four different oversize models, including "The Whitney," a metal casket with a 35-inch interior width.

Those who retrieve bodies have also noticed the changes.

Ernie Heffner of Heffner Funeral Homes said he more frequently needs to bring along extra people when transporting bodies to help with the lifting.

Paramedics and emergency medical technicians have also had to struggle increasingly with overweight patients and bodies.

Consequently, the number of back injuries industry-wide has been increasing, said David Gent, an EMT [emergency medical technician] with Red Lion Ambulance, located near York, who has suffered back pain from lifting people too heavy for him. They are required to ask for assistance for any patient weighing more than 250 pounds.

In order to "save our backs," Gent said, Red Lion recently purchased two new stretchers to accommodate greater weight. The stretchers use hydraulics, rather than muscle, to lift and lower patients.

But as others grapple with how to deal with the increasing size of the dead, Jack Sommer said he doesn't think it's yet affected the cemeteries. The manager of Prospect Hill Cemetery and Cremation Gardens in York said he deals with oversize caskets periodically. And occasionally, he needs to custom order a vault.

"I do about 300 burials a year," Sommer said. "But I can't say that I'm seeing an increase statistically in oversize caskets." Heffner said the reason is that the casket industry so far has been able to design models that still fit in a standard-size vault.

While caskets are getting roomier, the expansion design has focused, so far, on the interior—not unlike the auto industry redesigning a compact car for added legroom.

But if Americans keep putting on the weight, eventually cemeteries will need to redesign as well. Then, burial plots will have to get bigger, mausoleums will hold fewer bodies and "you won't get as many spaces in a family plot," Heffner said.

"If it's not impacting cemeteries yet," he said, "it will."

■ Rebuilding Wall Street

Accountants, in a Reversal, Say Costs from the Attack Aren't "Extraordinary"

STEVE LIESMAN

Wall Street Journal (Eastern Edition)   New York, N.Y.
October 1, 2001

In a decision sure to stir disbelief, a task force of the Financial Accounting Standards Board [FASB] has determined that the terrorist attack of Sept. 11 was not an "extraordinary" event—at least as the accounting world uses the term.

The FASB's Emerging Issues Task Force, after a debate infused with rare drama for the accounting world, decided unanimously Friday not to allow companies to treat costs and expenses related to the disaster as an "extraordinary item" in the financial statements they file with government regulators.

In other words, costs that companies consider attack-related won't be broken out as a separate line item below income from continuing operations but instead will be considered costs that are part of normal business operations. The ruling of the 13-member task force is effective immediately and should be published on its Web site (www.fasb.org) tomorrow.

"The task force understood this was an extraordinary event in the English-language sense of the word," said Chairman Tim Lucas. "But in the final analysis, we decided it wasn't going to improve the financial-reporting system to show it" as an extraordinary item standing separate from continuing operations.

The decision Friday marks a reversal of a tentative agreement announced earlier last week, in which the task force reached the opposite conclusion and had issued a 17-page draft statement that offered guidelines for how companies might account for the disaster as an extraordinary item.

But as the task force discussed how companies would actually apply the guidelines, the debate became so complex that members effectively threw up their hands. They found it impossible to separate the direct financial and economic effects of the terrorist attack from the prevailing economic conditions before the event and the impact on the economy afterwards.

The impact of the attack was so pervasive, affecting virtually every company and creating such a broadly new economic landscape, that "it almost made it ordinary," said task-force member Dick Stock.

The task force's decision represents one of the first installments of what is likely to be a lengthy debate over the economic impact of the terrorist attack. Some economists contend that it plunged a fragile economy into recession; others say the economy already was in recession and the attack only deepened the downturn.

The ruling could make it somewhat more difficult for investors to figure out how much impact the attack had on a given company, since the losses won't now appear in a single line on an income statement. But the ultimate decision suggests that investors should be wary of some company claims—as will be expressed in news releases and other earnings-related commentary—about what should and what shouldn't be attributed to the disaster.

Indeed, some FASB members voiced concern that companies would blame the attack for a variety of unrelated costs and expenses.

The FASB is a private-sector group charged with establishing generally accepted accounting principles [GAAP], which publicly traded companies must follow in the financial statements they file with regulators. During the past several years, however, companies and Wall Street analysts increasingly have moved away from GAAP accounting in their earnings news releases. They have instead presented earnings results in a variety of different ways that exclude many costs, such as restructuring charges or asset write-downs, that GAAP considers part of the ordinary course of business. The FASB holds no sway over what companies say in earnings releases.

Since many companies will likely use those releases to present their own findings about the disaster's impact, the most significant effect of task force's decision is in the official bookkeeping: Companies will be prohibited from showing losses related to the disaster as an extraordinary item in their official income statements. Companies, however, will be free to break out such costs in footnotes to their financial statements and in the discussion of management activities, which are part of the official financial statements.

The word "extraordinary" has a highly restricted use under GAAP. To qualify for such treatment, an item must be both unusual in nature and infrequent in occurrence. Losses from an earthquake in a city that has rarely experienced them would qualify. Yet losses from an earthquake in San Francisco wouldn't.

Mr. Lucas said that treating the event as extraordinary, in any case, wouldn't have captured all the losses from the attack. The largest "cost" for many companies is sure to be lost revenue. But the amount of lost revenue can never be accurately calculated, as a company would be hard-pressed to say exactly how much it would have taken in if not for the attack.

The task force's attempt to create guidelines for treating the attack as extraordinary at first proceeded smoothly. It quickly decided that the event wasn't extraordinary for insurance companies, as they are in the business of underwriting such events. While an attack of this size has never occurred before, said one of the members, "most of the things that happened have happened before."

Indeed, it was even mentioned during the debate that the 1993 terrorist bombing of the World Trade Center was at least one reason not to consider this attack extraordinary for insurers.

But when the task force moved on to the airline industry, the most explicitly affected industry, it quickly became clear that it couldn't create comprehensive guidelines for all companies.

The first and last snag in the debate involved how to account for losses to the value of assets not physically damaged in the attack. Members understood that airplanes had lost some unknown value as a result of the event. But how much value? The industry was slowing prior to Sept. 11. AMR Corp., the parent company of American Airlines, had warned in a Sept. 7 news release that it expected its third-quarter loss to be "considerably larger than its second-quarter loss as it continues to feel the combined effects of a weak economic climate, high fuel prices and increased labor costs."

It also had announced plans to retire older aircraft earlier than previously planned, suggesting it already was in the process of writing down aircraft values.

Once the task force found it couldn't tackle the airline issue, it quickly reached a reluctant consensus to prohibit all extraordinary treatment of the event.

In a letter to task force last week, United Technologies Corp. Controller David Nord maintained that while the event "likely accelerated and magnified the downturn, it is impossible to separate the extraordinary component from restructuring actions that were likely inevitable."

Ironically, Mr. Nord suggested that the end result will be greater clarity of financial statements. Since the process of separating the extraordinary from the ordinary would ultimately have been subjective, the bottom-line net-income figure calculated according to GAAP will remain an objective statement of a company's financial results. And since all companies have in some way been affected, and all companies must deduct losses from operating income, investors will still be able to compare earnings within and across industries.

When investors return to evaluating corporate performance, accountants suggested, they will be better served by a decision that has kept the playing field level for the vast majority of companies.

"We need to make sure that we keep the financials clean," Mr. Nord said.

DOUBLE STANDARDS

The History of Standardizing Humans
in Modern Life Insurance

Martin Lengwiler

The Historical Conditions of Standards for Humans

■ In this chapter, I examine the history of standards for humans in life insurance since the late nineteenth century. Life insurance is an exemplary institution within which to examine the definition and implementation of human standards in modern societies. Contemporary practices of standardizing humans have increasingly attracted interest in recent science studies. In a review article, Lawrence Busch (2000, 278–80) discusses eight forms of standards, three of them in relation to humans: standards addressing workers (e.g., technological standards in industrial production affecting work practices), standards affecting capitalists (mainly the organization of their means of production), and standards for consumers (through the standardization of consumer goods or of distribution systems). Geoffrey Bowker and Susan Leigh Star (1999, 165–225) analyze the effects of classifications, in more or less standardized forms, on the biographies of individuals, such as the consequences of medical diagnoses on the biographies of patients or of race classifications by the South African Apartheid regime on people with a racially mixed background. Several articles in this book similarly investigate the standardizing of several dimensions of human existence, such as bodily features, working or aging, and its social consequences (see Steven Epstein, chap. 2; Martha Lampland, chap. 5; and Judy Treas, chap. 3 in this volume).

This chapter owes much to the critical suggestions of two anonymous reviewers. It also profited from discussions with the participants of the group in residence at the Humanities Research Institute at University of California, Irvine, namely Geof Bowker, Steve Epstein, Rogers Hall, Martha Lampland, Jean Lave, Janice Neri, Ted Porter, L. K. Saunders, Leigh Star, and Judy Treas. Any remaining errors are of course entirely the author's responsibility.

Despite these sociological insights, little is known about the historical origins and development of standards for humans. The history of human standards is usually traced back to the history of statistics, for example to the mid- and late-nineteenth-century concepts of statistical means and normality in the influential works of Adolphe Quetelet and Francis Galton. Adolphe Quetelet (1796–1874), a Belgian statistician, designed a quantitative moral statistics, an approach drafted against the more qualitative sociology of August Comte. His notion of the average man had a powerful influence on Francis Galton's (1822–1911) works on the normal curve and his understanding of the physical characteristics of human beings as normally distributed. In the tradition of Quetelet's and Galton's work, the concept of a standard human stands for the normal or average human in the sense of a statistical mean, often interwoven with an idealistic emphasis of the mean segments of the normal distribution (Porter 1986, 40–70; see also Epstein, chap. 2 in this volume).

This common historical perspective, however, is often restricted to an epistemological analysis and to the history of ideas (e.g., Stigler 1986; Porter 2003). Therefore, by focusing on insurance, I aim in this chapter at expanding the epistemological view with a social and institutional approach (see also, Wagner 2003). Since their emergence in the nineteenth century, modern standards for humans were applied and redefined in institutional and practical contexts, such as modern state bureaucracies, administrations of welfare institutions and social insurance, medical institutions, and private corporations. Combining the epistemic and practical levels facilitates the analysis of the use of standards as an instrument of social power. The privileges of setting standards are intrinsically linked to social power relations in modern society. "The right to determine measures," as Witold Kula, Polish historian of science, points out, "is an attribute of authority in all advanced societies" (1986, 18). Similarly, Lawrence Busch is critical of the notion that standards are neutral because this ignores the "moral economy of grades and standards" in contemporary societies (2000, 275). As an instrument of authority, standards are used, for example, to distinguish between normal and abnormal (nonstandard humans) or between accepted universal and marginalized (local or traditional) forms of knowledge.

Some aspects of the institutional history of modern standards have already been explored. Well known is the standardization of intelligence around 1900 through the calculation of an intelligence quotient (IQ) in formalized intelligence tests. The widespread application of intelligence testing in schools or the army by psychologists as Alfred Binet, William Stern, or Lewis Terman was an important starting point for the rise of the category "feeblemindedness" in twentieth century psychology and psychiatry (Gould 1996, 204–54; Lengwiler 2003, 49–54). In a similar way, the historiography of the eugenics movement has shifted from epistemic to institutional issues, by showing how the theories of a normal inheritance (or genotype), as developed by Francis Galton or later by Wilhelm Johannsen, were used to legitimize the political program of the eugenics movement aiming at restrictive marriage legislation or the sterilization of women with supposed hereditary deficiencies (Dowbiggin 1997; Broberg 1996). Standardization practices have also been an issue in labor and social history, in studies linking the origins of Taylorism to late-nineteenth-century debates in physiology about the potentials of a "human motor" and its implications for the standard organization of the workplace (Rabinbach 1990; see also Lampland, chap. 5 in this volume).

Institutionally, I focus in this chapter on the context of life insurance and on the related commercial insurance business—an area largely unexplored in recent research. Why insurance? First, because life insurance plays a crucial role for the development of statistical notions of standard and normality. Authors such as Lorraine Daston (1988), Theodore Porter (1986, 2000), and Ian Hacking (1990) point out that the technical calculations used for life insurance were an important experimental field for models of modern statistical and probabilistic thinking. Second, and more generally, insurance can be seen as an exemplary social technology of modern societies, following, for example, the arguments of François Ewald in his study of the history of accident insurance in France. Ewald highlights the increasing social relevance of insurance since the late nineteenth century, molding this process in a Foucauldian tradition as an *assurantialisation* (an "insurancizing") of modern societies, eventually producing, in Ewald's terms, an "insurance society" (Ewald 1986, 389–95). Life insurance, in particular, witnessed a long-term expansion from its origins as a form of providence limited to the self-employed middle classes of the eighteenth and nineteenth centuries to a form or protection that during the twentieth century became prevalent in the middle classes and increasingly also in the working classes (Clark 1999, 155–69; Raynes 1964, 366–91; Arps 1970, 480–82). The third reason for analyzing life insurance is that the business of commercial life insurance was fundamentally marked throughout the twentieth century by the distinction between standard and substandard lives—a conceptual and practical distinction that is at the core of this chapter.

Thus, the debate about insuring substandard lives serves as an exemplary revealing case to examine the ambivalent practical effects of modern human standards, between exclusive and inclusive, discriminating and privileging, and disabling and enabling practices. In the first section, I analyze the historical conditions that gave rise to the notion of standard lives in late-nineteenth-century insurance, pointing out the cultural, economic, and professional factors behind the standard-substandard distinction in life insurance. In the second section, I examine the gradual transformation from originally prohibitive to more inclusive insurance standards and the differentiation of insurance classifications related to this process. This differentiation gave rise to the more sophisticated standards and classifications in life insurance examined in the third section. In the fourth section, I show how discussions about substandard risks have influenced the practices of life insurance medicine until the recent past. I end the chapter with a summary and some concluding remarks. The analysis is based on scholarly debates among American and European actuaries and insurance physicians since the late nineteenth century. Thus, the study is embedded in a comparative framework that juxtaposes the cases of American and European insurance.

Cultural and Economic Conditions
for the Emergence of Human Standards

The distinction between standard and substandard lives is not just a marginal note but one of the central themes in the history of modern life insurance. Before the 1880s, the concepts of substandard (or abnormal or nonstandard) lives did not exist in insurance, at least not as formalized concepts and standardized practices. Instead, medical examinations in life insurance varied from company to company. Most life insurance companies followed the principle that their commercial performance did not so much depend on the minute calculation of the customers' risks as on a restrictive selection of their clientele (Porter 2000, 227–29). The main task of insurance physicians was, therefore, to distinguish between healthy and unhealthy applicants and to exclude the latter from access to insurance policies (Davis 1981; Sturm 1935, 29). Most companies had their own examination rules, and contemporary observers such as the medical director of the American Prudential characterized the situation as "chaotic" (Brown 1935, 94). Usually, medical examinations consisted of a few basic diagnostic checks, mainly focused on tuberculosis, the most important cause of death in younger people and therefore the disease feared most by insurance companies (Davis 1981, 398; Sturm 1935, 57). Insurance physicians also collected information on the family's medical record and the applicants' individual history of diseases. The use of simple diagnostic instruments, such as the stethoscope, was common, especially to detect early stages of tuberculosis (Davis 1981, 395–98; Brown 1935, 94). Applicants with tuberculosis were usually turned down, but the occurrence of the disease in the family occasionally allowed for restricted forms of insurance contracts. There are plenty of national specificities. The American insurance business, for example, preferred limited endowment contracts for applicants with a family history of tuberculosis, guaranteeing, similar to a pension, a regular payment after a fixed number of years. British companies favored adding a certain number of years to the applicant's age (Brown 1935, 94; Greene 1905, 347, 377–83).

Since the 1890s, the professional circles of insurance doctors and actuaries increasingly debated the homogenization and standardization of the disparate insurance practices. This debate, focusing on a common definition of standard and substandard lives, quickly became the most important issue in actuarial theory and life insurance medicine. By the interwar period of the twentieth century, leading experts deemed the definition of standard and substandard lives the most important classification for the development of modern life insurance practices

(Florschütz 1925, 48–50). The concept of standard lives is based on the notion of a Galtonian normal distribution, also addressing the technical needs of commercial insurers to define a uniform premium policy. In this sense, the notion *standard* referred to an insurance client with health risks considered to be normal. Standard lives were insured under the common conditions, paying the standard premium and belonging to the same risk pool as the other regular clients. The notion of substandard lives (synonymous terms include abnormal, under-average, and aggravated lives) referred to clients with ailments or risks deemed extraordinary and elevated; they were either insured with a surcharge or—originally the most common practice—completely excluded from access to life insurance (Hörnig 1935b, 8; Brown 1935, 94). The concepts of standard and substandard thus marked a threshold and an entry point after which a client was granted the regular (i.e., standard) form of insurance services. The insurance concept *standard* meant normal in both a statistical sense (a risk deemed normal) and in a normative sense (a customer granted access to life insurance).

Indirectly, the distinction between standard and substandard risks mirrors the degree of solidarity that an insurance company stipulates among its clients. If a company insured all its clients at the same rate, irrespective of differences between individual risks, the solidarity between the clients and the balance between the risks was high. If, instead, a company made a difference between standard and substandard risks, either by excluding substandard cases or by charging additional premiums, the solidarity between the clients decreased.[1] Thus, the definition of a *standard life* also reflects the degree of solidarity in life insurance.

What ailments were treated as substandard risks? In the late nineteenth and early twentieth centuries, the two diseases most commonly identified with substandard lives, amounting to more than 50 percent of the cases, were tuberculosis and heart diseases (including angina pectoris and "heart rheumatism"; Gruder 1927, 299). Similarly treated as substandard cases were applicants with cancer, progressive paralysis (syphilis), mental illnesses, and diabetes, not to mention applicants with disabilities (Sturm 1935, 51–59). The overall proportion of applications excluded as substandard cases was more than 10 percent. According to a Belgian statistical analysis, from 1863 to 1893 insurance companies in Germany, Austria, and Denmark rejected between 12 and 29 percent of their applicants, with an average of 18 percent (Poëls 1927, 278). In Germany, the quota of denied applicants was even higher, between 22.7 and 23.1 percent in 1883–1893, or

---

1. Today, actuaries speak of the difference between the solidarity principle and the equivalence principle. The *equivalence principle* applies to insurance schemes in which clients are insured after their specific, individual risks—with a premium equivalent to their risk. This is also the reason why the concept of standard risk applies only to the sector of private insurance and not to social insurance. Social insurance, such as workmen's compensation or compulsory old-age and health insurance (as in European welfare states), insures their members within one risk pool, which is one of the principles of social insurance.

between 18,000 and 27,000 applicants per year. Taking into account that these figures include multiple applications by the same person, it is still a reasonable guess that, around 1900, up to 20 percent of applicants for life insurance were turned down as substandard risks (Gebauer 1895, 293). These figures are a reminder that behind the technical debate about the distinction between standard and substandard risks there were several generations of chronically ill and disabled people who suffered from systematic stigmatization and discrimination by life insurance companies.

How did the distinction between standard and substandard lives come about? Two historical factors led to the emergence and formalization of the distinction around 1900: (1) widespread cultural fears about the decline and degeneration of modern societies and (2) the economic constraints of a growing mass market for life insurance.

The relation between the cultural pessimism of the turn of the century and the debate about substandard lives has two dimensions: a medical one and a semantic one. On the medical level, the discussion about substandard lives emerged as part of the growing social and cultural concerns about the supposed spread of hereditary diseases, a concern that was also epitomized by the popularization of degeneration theory. In this sense, the debates about substandard lives and about degeneration theory are parallel and interrelated phenomena. Until the 1880s, before the rise of the notion of substandard risks, commercial insurers used a variety of heuristic procedures to deal with unhealthy clients. In mid-nineteenth-century Germany, for example, the earliest life insurance referred to general medical concepts such as the notion of the *Ungesunden* (the unhealthy) to categorize applicants to be excluded from insurance policies. For example, Ludwig Elster, a political economist and an adherent of the influential left wing of the German Historical School, discussed the topic in his influential work on "Life Insurance in Germany" under the heading "Versicherung Ungesunder" (insurance of unhealthy people) (Elster 1880, 18; Gebauer 1895, 290; Grimmer-Solem 2003, 278). The chief medical officer of Basler Life Insurance wrote of *fehlender Widerstandskraft* (lacking resistance) or *Widerstandsschwäche* (a feeble resistance) to diseases (Haegler 1896, 1–6). The concept of degeneration was also missing from the often-cited compendium of Eduard Bucheim, Austrian insurance physician, on the "Medical Diagnostics in Insurance," published in 1887. Moreover, before 1900 clients were not only excluded for medical reasons but also for particularly hazardous occupations or dwelling places (Kehm 1897, 71).

In the 1890s, however, the debate about medical criteria for life insurance was intensified by the growing fears of an increase of hereditary diseases, which paralleled the first comprehensive definitions of substandard risks. The German term for substandard lives, *minderwertige Leben*, for example, was at the core of two scholarly treatises of 1895 and 1897, both published in the series of Studies in Political Economy edited by Ludwig Elster (Kehm 1897, 1; Gebauer 1895, 290–93).

Max Gebauer wrote his study as a dissertation with Elster; Max Kehm was another political economist close to the German Historical School and wrote specifically about the insurance of substandard risks. Both were concerned with the difficulties of insuring a growing number of negative hereditary dispositions and debated whether supposed (but not yet manifest) diseases should be treated as substandard risks and to what extent these cases should have an elevated premium imposed on them (see the discussion of Gebauer's position in Kehm 1897, 74–83). Other contemporary insurance experts, medical as well as actuarial, shared the concerns of Gebauer and Kehm. Two Austrian life insurance doctors, Ernst Blaschke and Eduard Buchheim, also pointed at the difficulties of insuring customers with inherited illnesses. In a hereditary model, not only the diseases of the applicant but also the medical history of the family became relevant for the medical decision about whether to insure. Both authors agreed that an applicant could be categorized as substandard even when he or she was not actually ill; it was enough that one of the siblings or parents was affected by a diseases seen to be hereditary, for example, tuberculosis, cancer, gout, rheumatism, heart diseases, mental illnesses, and neurological diseases (Buchheim 1887, 11–16; Blaschke 1895, 22–25).

The cultural background for these growing concerns about hereditary diseases was characterized by the broad diffusion of degeneration theory in the final decades of the nineteenth century. Degeneration theory originated in the works of Benedict Augustin Morel (1809–1873), French psychiatrist. Influenced by Georges-Louis Buffon's natural history and Henri-Marie de Blainville's zoology, Morel drafted a general theory of the evolution of inherited illnesses, which he published 1857 in his "Traité des Dégénérescences." Morel was mainly thinking about mental illnesses. His main thesis was that mental illnesses were hereditary and marked by progressive deterioration over the generations, a process that he called degeneration. Morel also described a series of mostly physical "signs of degeneration"—from the small size of the skull to the accrued earlap—which he used to diagnose degeneration (Pick 1989, 44–50; Schmuhl 1987, 59–63; Dowbiggin 1985, 205; Lengwiler 2000, 111). In Germany, Morel's theory of degeneration was first received in psychiatry, but two decades after the death of the theory's founder. The most influential German adaptation of Morel's theory was written by the Julius Ludwig August Koch (1841–1908), an asylum director of the state of Wurttemberg. His three-volume treatise on "*Die psychopathischen Minderwertigkeiten* (the psychopathic degenerations) was published in 1891–1893 and found an audience among the most notable psychiatrists of the time, such as Emil Kraepelin, Richard von Krafft-Ebing, and Theodor Ziehen. The theory of the psychopathic personality that Kraepelin derived from Koch's concept of psychopathic degenerations became the starting point of modern psychiatric theories of personality disorders (Koch 1891–1893; Weingart, Kroll, Bayertz 1988, 47–50). Following Koch's terminology, medical sciences in Germany translated the term *degeneration* either directly into *Degeneration* or into the Germanized word *Entartung*, whereas *degenerate* was usually translated *entartet* or *minderwertig*. By 1900, degeneration theory spread to other disciplines, both to other medical subdisciplines and to nonmedical disciplines such as anthropology and political economy (Pick 1989; Weingart, Kroll, Bayertz 1988, 302–4, 474).

The relation between the fin-de-siècle cultural pessimism, epitomized by theories of physical degeneration and social decline, and the rising debate about substandard lives is also illustrated by a semantic link. The German concept *Minderwertigkeit* was used for both the notion of degeneration (in Morel's sense) and the concept of substandard lives in the insurance industry. Until the 1880s, the term *Minderwert* (literally, "reduced value") was mainly used for changing property values (Grimm and Grimm 1885, 2231). Only within the medical degeneration discourse did *Minderwertigkeit* occupy the same semantic space as *minderwertig* in insurance medicine (e.g., Rudolph 1927). By the interwar period of the twentieth century, the concept had gained its discriminatory meaning that was eventually picked up and widely used by the Nazi programs for the persecution, sterilization, and assassination of the physically and mentally disabled (Schmuhl 1987, 79–86, 94, 261–90; Weingart, Kroll, Bayertz 1988, 302–4, 474).[2] Thus, the actuarial and medical debate about substandard risks linked the notion of standards to physical normality and abnormality and thus helped to prepare the semantic ground on which the national socialist terminology was built.[3]

Apart from these cultural reasons, the standardization of examination practices in life insurance also made sense economically. In most Western countries around 1900, life insurance was about to enter the mass market. Between 1870 and 1890, the accumulated value of life insurance policies doubled in the more mature markets of Britain and the United States, and quadrupled in the still-emerging markets of France and Germany (Arps 1965, 142). Many actuaries complained that under these conditions the disparate procedures of the medical examinations were a severe economic handicap. Charles Lyman Greene, in a British textbook on life insurance medicine published in 1905, criticized the many ways that commercial insurers defined the criteria for substandard risks. Greene complained that there was no "thoroughly sound, safe, and precise method" of estimating substandard lives and that the subject was involved in a "maze of contradictory opinions and theoretic considerations." His advice was pragmatic. He encouraged companies to start insuring substandard lives; the subsequent economic pressures of the mass

2. For the semantic field of *Minderwertigkeit*, see, for example, *Meyers Lexikon*: "Mindigwertigkeit: quality of a thing or a person relative to the average of the good and the worthy. . . . Minderwertig are those members of a people or a kin that are asocial, unfriendly, or hostile to a community, for example, in terms of their ability and character (which are mostly inherited)" (1939, vol. 7, 1418).

3. After 1945, the term *Minderwertigkeit*, because of its association with the national socialist persecution of the disabled, was abandoned in life insurance terminology and replaced by the technical concept *aggravierte* (aggravated) risks (Stévenin 1968).

market would then make it "necessary to consider just what is meant by a standard life" (Greene 1905, 342). In this sense, it was no coincidence that the emerging mass market in life insurance was paralleled by the clarification of the distinction between standard and substandard risks. The economies of scale of capitalist mass production acted as powerful leverage for the standardization of insurance practices.

Professional Competition: From Prohibitive to Discriminatory Standards after 1900

The development of a clear definition of substandard risks and of technically sound criteria for whether to insure such cases was a complex process with several stages. Crucially, in the course of this process the practical implications of being labeled a substandard risk changed considerably. Whereas until the turn of the century substandard cases were usually excluded from access to a policy, after 1900 substandard risks were gradually included in life insurance, mainly by adding a surcharge to the premium. Before the First World War, several major American and European insurance companies replaced their prohibitive policies with more inclusive insurance schemes, appealing to previously rejected groups of customers. Although at first these were just scattered tendencies, the inclusion of substandard lives became a common trend in Europe after the war. By the admission of previously rejected people, the effect of insurance standards gradually changed. Previously prohibitive standards, distinguishing only between insurable and uninsurable risks, now gave rise to more differentiated, although still financially discriminatory classifications of insurance customers. This process was enforced by the development of life insurance medicine that investigated the different life expectancies of substandard risks. Another aspect of this process was the development of standardized tests, shared by some of the major insurance companies, used to distinguish normal and abnormal risks. Thus, the insuring of substandard risks illustrates the ambivalent effects of standardization processes. Epistemologically, the measurement of substandard risks was first used to legitimate exclusive insurance schemes, but it eventually enabled the precise calculation, and thus the insuring, of previously excluded risks.

The transformation from prohibitive into classificatory standards was neither quick nor comprehensive. Until 1900, the impact of the theoretical debate on business practices in life insurance was limited—the insuring of substandard risks remained a marginal phenomenon (Gebauer 1895, 292). Still, there were several important initiatives already in the nineteenth century. As early as 1824, the British Clerical, Medical and General Life Assurance Company decided to insure all previously excluded applicants by setting a surcharge. The company had no statistical experiences, but tried to fix the amount of the surcharge according to the medical examination of its physician. Between 1834 and 1843, the Clerical counted 1,297 new customers under this rule. The average mortality among the group of abnormally insured clients was 50 percent above the mortality of regular insurance clients. Still, the company deemed its new insurance scheme a success. After 1850, the policy to include abnormal risks spread in the British insurance market. By 1872, two-thirds of all insurance companies had introduced similar offers (Kehm 1897, 65–71; Gebauer 1895, 290; Hörnig 1935c, 577).

In the 1860s, the British experiences were taken up abroad, namely in the United States and Germany, although with mixed success. The American Universal Life, for example, started to insure substandard lives after 1865 by offering contracts with reduced insurance benefits. But the business was not profitable enough, and the company stopped its venture in 1877 (Hörnig 1935c, 577). The first German company to offer insurance for substandard lives was the Friedrich-Wilhelm in 1867. Apparently, its experience was encouraging because in 1878 the first German insurance company was founded that was exclusively concerned with previously rejected customers. Unfortunately, this new Allgemeine Lebensversicherungsanstalt zu Leipzig A.G. did not live up to its expectations and went out of business after only three years (Hörnig 1935c, 578; Gebauer 1895, 291).

What were the reasons behind the trend toward more inclusive insurance schemes? One important reason relates to the establishment of social insurance in Germany and other European countries since the 1880s. Social insurances offered comprehensive schemes, insuring entire social groups such as industrial workers and old people. With social insurances, the individual risk assessment, such as the distinction between standard and substandard risks, was not an issue. On the contrary, social insurances were designed to appeal to the lower and poorer classes, which could not afford life insurance; the institution of social insurance intended to protect those parts of society left uncovered by private insurance. In Germany, Austria, and Switzerland, for example, the social insurance legislation after 1883 established national institutions for accident insurance, thus nationalizing a part of the formerly private insurance economy (Ritter 1998). Some contemporary voices went even further. In order to resolve the problem of people excluded from being insured, Adolph Wagner, one of the so-called *Kathedersozialisten* (socially sensitive liberal economists partly organized in the Verein für Socialpolitik), even argued for the nationalization of the whole private insurance sector (Gebauer 1895, 297–300; Wagner 1881). Arguments for the nationalization of the private insurance market were also uttered in the debate on substandard lives. Max Kehm (1897, 1, 72–75), political economist, for example, demanded a special public insurance scheme for substandard risks.

The institution of social insurance provoked a competitive reaction among the private insurance companies. The answer was to reach out to new customers and expand the business. Before the mid-nineteenth century, life insurance appealed almost exclusively to parts of the bourgeoisie, namely to traders and manufacturers, public servants and Protestant priests (Clark 1999, 155–69). Only in the latter part of the century did life insurance companies begin to address a mass market. In the 1890s, after Germany and Austria had introduced their first social insurance, several private insurance companies, mainly in German-speaking countries, took up their social appeal and started to promote the idea of a *Volksversicherungen* (people's insurance), a cheap (in every sense) life insurance for so-far uninsured groups, mainly the working class (Lexis 1906, 21). At the beginning, the people's insurance was a quite successful venture. Between 1902 and 1910, the proportion of the population in Germany that had life insurance rose from 10 to 20 percent, mainly due to the attraction of the people's insurance (Manes 1913, 42). In the interwar period, however, most people's insurance policies in Germany failed during the inflation crisis in 1922–1923 and the subsequent collapse of the insurance market (Lexis 1906, 21; Lederer 1914). The initiatives to insure substandard risks followed the same strategy—expand private insurance schemes to the lower classes and thus counter the political arguments for the nationalization of private insurance (Kehm 1897, 1).

The second reason for insuring substandard risks has to do with the economics of insurance itself. Insurance companies, in order to distribute their risks, tended to accumulate as many calculable risks as possible. Thus, the inclusion of new previously uninsured risks was also a matter of self-interest for the insurance companies. In the case of substandard risks, the problem was not so much that the risks were elevated but that the necessary surplus on the premium was hard to calculate. Because substandard lives were usually rejected, the insurance companies lacked the necessary statistics. The task was to find a practical solution for statistically investigating the prognoses of these rejected cases (Kehm 1897, 8–63). Obviously, large companies, Europeans and especially Americans, with their existing actuarial personnel and their statistical data, were in a favorable position to investigate substandard risks.

The actuarial background of most calculations of substandard risks points to another important element in the diffusion of the standard-substandard distinction—the competition between the profession of life insurance physicians (the more traditional experts in life insurance) and the actuarial experts who gained their professional status in commercial insurance only in the late nineteenth century. By designing standard measures to calculate substandard lives, actuaries tried to promote their own expertise and thwart the established status of their medical colleagues.

There is no better case to illustrate this professional competition between actuaries and insurance physicians than the history of the numerical method, developed by the New York Life Insurance Company after 1890 and one of the earliest and certainly the most influential initiatives for calculating substandard risks (Porter 2000, 234–39). The numerical method was by far the most influential actuarial tool for assessing substandard risks; it was widely adopted after 1900 in the United States and Europe. It was designed by Oscar Rogers, a New York Life physician, and Arthur Hunter, his actuarial colleague. They started their work in the early 1890s, reviewing internal statements of insurance agents complaining that with the current medical examinations too many good risks were unnecessarily declined. The problem for Rogers and Hunter was that all the accessible data were on the accepted insurance contracts, but, to find an answer for these complaints, they needed information on the refused applicants. Hunter suggested looking at the files of the rejected risks. In the end, a research project based on 25,000 cases was tied together (Brown 1935, 94–98).

The project lasted over nearly a decade, and its main conclusion was that a large number of the declined applicants could have been insured without any further risks for the company. As a result of the inquiry, the authors suggested a new technique to determine substandard risks. Based on a statistical analysis, Rogers and Hunter defined a list of nine "factors determining the risk" (i.e., influencing the mortality of their cases): build (weight in relation to height and age), medical history of the family, occupation, personal medical history, habits (e.g., use of alcohol and drugs), physical condition (i.e., impairments), habitat, moral hazard (e.g., reputation and financial situation), and the insurance plan requested by the applicant (because mortality was particularly high among people asking for short-term, speculative insurance contracts) (Brown 1935, 94–98). The inconsistencies in the list show that the factors were mainly derived for practical purposes (to be measurable), without a coherent analytical framework. Rogers and Hunter also developed a method to calculate and sum up these nine factors. The results of each risk factor were translated into a number (10 percent, 20 percent, etc.) either to be added to (in case of special risks) or deducted from (in case of qualifications reducing the risk, such as a secure occupation) a normal 100% premium. At the end of the calculation, all the additions and deductions were summed up, resulting in a final surcharge to the insurance premium. Thus, architects, teachers, and banking clerks had a comparably small risk factor of $MI10, whereas fire fighters, stonemasons, and saloon owners were measured with an increased rate of $PL50 (Brown 1935, 94).

Eventually, New York Life introduced the numerical method in 1906, both in the United States and in its European branches. The success of the numerical method represents, as Ted Porter has stressed, a crucial step in the revaluation of actuarial against medical knowledge. Before the innovation of Rogers and Hunter, the New York Life decision about whether to accept an insurance application was determined by the insurance physicians; with the numerical method, however, the medical diagnoses were reduced to a few simple questions about the applicant's general physical condition and the rest of the decision was transferred to the actuarial calculation of risk factors (Porter 2000, 234–39). Unsurprisingly, the opposition to the numerical method came mainly from insurance physicians, who feared the deskilling of their expertise. In Europe, the medical profession eventually succeeded, at least partially, in countering the actuarial innovations with their own medical form of standardized measurement.

Applying Double Standards?: Class-Based versus Individualizing Classifications

After 1905, the insuring of substandard risks was adopted by several European countries, partly inspired by the example of the European branches of New York Life. In the nationally specific insurance markets in Europe, two different institutional models evolved. The first model founded a new national insurance company dealing only with substandard risks. Usually, the initiative to establish such companies came from the life insurance companies, which intended to share their substandard policies in a common risk pool. This model was first adopted in Holland (by De Hoop in 1905), followed by Sweden (Sverige, 1914), Denmark (Dana, 1916), Germany (Hilfe, 1916), Austria (Verband zur Versicherung anormaler Leben, 1916), Norway (Norske Folk, 1917), Finland (Warma, 1920), Czechoslovakia (1922), and Italy (1928) (Palme 1927, 159–60).

But most of these companies remained unpopular among their potential clients. A contract with a specialized company automatically stigmatized the client as a substandard risk. The German Hilfe, for example, was already battling before the inflation crisis of 1922 against decreasing membership numbers. When the regular insurance companies started to offer insurance for substandard risks, offering their clients a nonstigmatizing form of contract, the Hilfe lost most of its customers to its competitors before it finally collapsed in 1922 (Rudolph 1927, 327).

The second, more popular institutional model relied on the already existing insurance companies, specifically on a collaboration between the life insurance and the reinsurance companies.[4] Contracts were signed between life insurance companies and their customers, but were managed by the reinsurance companies, with the advantages of a better distribution of the risk and of the opportunity to establish large statistical databases. The second model was adopted by Switzerland (Swiss Reinsurance) and Germany (Munich Reinsurance), both before the First World War (Sturm 1935, 78; Hörnig 1935c, 577–81; Florschütz 1925, 50).

Apart from these institutional models, two classification practices evolved in Europe. One was the numerical method, which by 1934 had achieved an officially recognized status as the standard procedure recommended by the Third International Conference for Substandard Risks in Prague (Coppet 1935, 106–9). The numerical system was used mainly in France, where it had been introduced by the French branch of the New York Life, and in Germany and Switzerland, where it had been introduced by the German and Swiss reinsurance companies (Coppet 1935, 106–9; Sturm 1935, 78; Rudolph 1927, 327). The other method, also widely used, especially in the Scandinavian countries, was based on risk classes. The original system of

---

4. *Reinsurance companies* are companies that specialize in insuring the risk of insurance companies. That is, their customers are other insurance companies and not individual clients, as in the case of primary insurance.

risk classes was developed in the 1890s by the Austrian insurance physician Eduard Buchheim. It is particularly interesting to compare these two classification models because they represent not only two different forms of professional expertise but also two different models of solidarity among the insured clients.

In Europe, the numerical method was often criticized as an individualizing classification, measuring the specific risk and fixing a particular premium for each applicant. This tendency provoked a widespread opposition by life insurance physicians, notably in Austria, Belgium, and the Scandinavian countries. The argument against the numerical method was that by focusing only on the individual it undermined the solidarity principles of insurance. In 1927, Otto Gruder, the secretary of the Austrian Insurance Association for Abnormal Lives, criticized the numerical method, saying that it would introduce a double standard into insurance practice. He argued that, when dealing with normal lives, life insurance had to abstract from the specificities of the individual case. The numerical method, he argued, contradicted the principles of insurance because it treated the substandard cases by individualizing them and not subsuming them into a homogeneous risk pool like the normally insured clients, for whom age was the only criterion to calculate the risk. It was this specific difference in the treatment of standard and substandard cases that Gruder criticized as a double standard (Gruder 1927, 288). Some opponents went even further. E. Poëls, a Belgian insurance physician and one of the founders of the International Associations of Insurance Physicians, and Louis Maingie, an actuary and his collaborator, argued for insuring substandard risks using the same procedure as standard risks. Examining the fate of 1,300 applicants of the Compagnie Belge d'Assurances Générales sur la Vie who had been rejected between 1893 and 1921, they concluded that several substandard risks were, in fact, not special risks at all because their mortality rate was similar to that of standard cases. This included risks such as "hereditary constitution," "weak constitution," "beginning artheriosclerosis {sic}," and "non-evolving syphilis"— risks that would have caused an extra premium to be charged using the numerical method (Stévenin 1968, 319; Maingie 1927, 275; Poëls 1927).

The alternative was a system based on a limited number of internally homogenous risk classes. The most important model for these risk classes was Eduard Buchheim's classification. Buchheim had worked since the 1870s as the head physician at the Erster Allgemeiner Beamten-Verein der österreichisch-ungarischen Monarchie (the Austrian government's pension insurance company) (Hörnig 1935c, 579; Buchheim 1878; Blaschke 1895).[5] His classification was originally based upon three risk classes: (1) hereditary diseases, (2) latent diseases or abnor-

5. As this shows, social insurance companies, such as Buchheim's pension insurance, were also interested in examining substandard risks, even though this distinction was not an issue for their insurance schemes. Because social insurance included all risks, including the substandard ones, among its clients, their records offered much more specific knowledge on the life expectancy of substandard risks.

mal conditions, and (3) acute diseases with mortality risk (mainly chronic diseases). Buchheim's model was more inclusive than the numerical method. Particularly in regards to his third class, he suggested insuring some illnesses that would be excluded under the numerical method, such as diabetes and minor cases of epilepsy (Karup, Gollmer and Florschütz 1902, 501–11). With only three classes, Buchheim's system relied on only three premiums; this reflected a comparably strong solidarity principle, certainly stronger than the numerical method.

However, Buchheim's model soon turned out to be too rigid. The Austrian Insurance for Abnormal Lives chose to use an adapted form of Buchheim's classification, enlarged to six risk classes in 1919 and a few years later to eight risk classes. The classes were defined by their respective proportion of extra mortality compared with normal mortality (class 1: excess mortality of 50 percent; class 2: excess mortality of 75 percent; etc). The medical examination was not intended to measure specifically the possible hazards and diseases of the applicant, as one under the numerical method did. The class model related each disease to one of the classes, and the insurance physician simply had to find the main diagnosis, which then determined the risk class (Gruder 1927, 288–90). Scandinavian countries such as Denmark and Norway adopted similar schemes with disease-related classifications (Drachmann 1927, 121–23). The Norwegian example also shows the limits of disease-based classifications. A small number of the disease-related classes did not allow for a differentiation according to the severity of diseases (e.g., heart diseases). In 1925, the Norwegians abandoned their classification based on diseases and switched to a system based primarily on extra mortality risk. But the risks were still grouped in six classes, in opposition to an individualized system using the numerical method (Aabakken 1927, 143; Hörnig 1935a, 112).

The Norwegian system also epitomizes a general trend in the way European countries developed classifications for substandard risks. By the 1930s, most countries used a mixture of the two opposing classification methods (the numerical method based on the calculation of individualized risk factors and a system of limited classes based on a set of diseases or disease groups). The disadvantages of both systems had to do with ideas of justice; the numerical method, with its focus on the individual, did not live up to the social idea of insurance (the balancing of risks according to the solidarity principle), whereas the classification by diseases was too general, squeezing together unjustly too many diverse risks into one group. The conciliatory solution was a differentiated classification, based on an extended number of classes, like in the Norwegian system.

Development of Twentieth-Century Life Insurance Medicine and the Persistence of Substandard Risks

The development of life insurance medicine is closely related to the insuring of substandard risks, mirroring the medical answers to the actuarial calculations. The early era of European life insurance medicine, between 1890 and 1930, was partly stimulated by American life insurance medicine and mainly dedicated to the collection and analysis of statistical evidence (Gruder 1927, 283). International collaborations had already been established before 1900, with Germany and Belgium as particularly active nations (in 1899, the Congress of the International Association of Insurance Physicians, later the International Congress for Life Insurance Medicine, was established; Stévenin 1968, 317–19; Florschütz 1925, 48; Brown 1935, 94), although the First World War and the following economic depression partly impeded further activities. Most of the international activities were suspended in the 1910s and early 1920s (Stévenin 1968, 323–25).

Starting in the 1920s, research in European life insurance medicine focused mainly on special substandard risks, such as tuberculosis, heart disease, hypertension, cancer, diabetes, and kidney diseases (for the United States, see Porter 2000, 240–43). Because the prognosis was not addressed in the early methods of classification, as in the numerical method, prognosis became one of the key topics in life insurance medicine. Some insurance physicians saw the general significance of their discipline in the contribution of prognostic theories to medicine (Stévenin 1968, 326; Florschütz 1925, 48). In 1934, the prognostic research was permanently institutionalized by the International Congress of Life Insurance Medicine in Prague. The congress decided to create a permanent international committee to unify the statistical data and coordinate the medical research. The Coopération Internationale pour les Assurances sur la Vie des Risques Aggravés (Cointra) still exists today and includes members from Austria, the Czech Republic, Denmark, Finland, France, Germany, Italy, Norway, Switzerland, and Sweden (Stévenin 1968, 325).

The long-term effect of research in life insurance medicine was that, by the 1960s, all previously excluded illnesses had been turned into insurable risks— although most often still with a surcharge on the premium (Griebler 1992, 26). The classification of substandard risks was thus reflected in an array of life insurance products. But insuring substandard risks did not mean that the discriminatory classification was in any way simplified or even dropped. Up to today, traces of degeneration theory are still discernible in life insurance medicine, for example, in the notion of a "hereditary disposition" that still counts as a characteristic for substandard risks (Griebler 1992, 28).

The history of Swiss life insurance offers an illustrative example of the persistence of the concept of substandard risks. In Switzerland, life insurance companies had already established a centralized database for all customers with substandard risks in 1891, a system that remained unchanged until 1992. The database was managed by the professional association, the Association of Swiss Life Insurance Companies. All associated companies were asked to contribute at least the names of applicants and policy-holders with "substandard risks" to what was called the Notification Pool for Abnormal Risks (Mitteilungsverband für abnormale

Risiken). All data in the pool were accessible to any company associated but hidden from the knowledge of the customers. As soon as an insurance company was approached by a new applicant, it checked the records of the notification pool, usually without telling the applicant about the inquiry. If the applicant was registered as an "abnormal risk," the company would either turn the application down or grant a restricted policy. Usually, the applicants did not learn what had led to the insurance company's decision. The hope of the Association of Swiss Life Insurance Companies was that its notification pool would prevent customers with substandard risks from cheating the insurance company by hiding their illnesses. Often, the data that the insurance companies delivered for the centralized database included not only the applicant's name and the decision of the insurance company but also information on the reasons for the decisions or even the complete medical record of the applicant.[6]

The notification pool was practically undisputed until the 1980s, mainly because hardly anybody outside the insurance industry knew about it. The reason why it finally became a controversial issue was the introduction of a Swiss law concerning data protection in 1992. After this legislation, the notification pool was no longer allowed to collect and handle data without the knowledge of the data subjects (the insurance clients). With the introduction of the law, the legal departments of major insurers became skeptical about the legality of the notification system. Finally, one of the market leaders pulled out of the system, causing a domino effect. As two other major insurers joined the first and dropped out, the whole notification system collapsed. The old hidden system was replaced by a new information pool, based on the consent of the customers. Nowadays, information on physical ailments is still regularly checked and fed into a common database of the Swiss life insurance industry but only after the approval of the applicant (Winterthur 2000).

Standards for Humans and Their Ambivalent Effects

The case of life insurance illustrates some crucial conditions for the emergence of standards for modern humans in the late nineteenth and early twentieth centuries. The standardization of the human lives to be covered by life insurance, including the standardization of the medical examination practices and the formalization of the distinction between standard and substandard lives, was due to the cultural, economic, and social transformations of modern industrialized societies. Culturally, the concern about substandard risks, in particular about the seemingly rising incidence of hereditary diseases, was part of a general form of cultural pessimism

---

6. Archives Zurich Financial Services, Switzerland, file Q 123 204: 42417, documents of 19 April 1994; 3 May 1994.

among the fin-de-siècle societies and thus related to other cultural models of decline, such as late-nineteenth-century degeneration theory. Cultural and social historians such as Daniel Pick and Ian Dowbiggin stress that these models of decline reflect the social tensions of modernizing societies in the late nineteenth century, such as the growing social diversity due to the transformation from agrarian into industrial societies (Pick 1989, 40) or—in the case of the rise of degeneration theory in France—the self-conception of a politically disappointed French bourgeoisie during the Third Republic (Dowbiggin 1991). In a similar sense, the distinction between standard and substandard risks in life insurance was originally designed to secure the institution of insurance and its healthy clients from the unhealthy parts of a degenerating society.

Economically, the standardization of insurance assessments reflected the growing mass market in life insurance at the turn of the century. The economies of scale of the insurance market demanded simplified, standardized products—a requirement that was met by the formalized actuarial assessments developed after 1900. Socially, the professional competition and rivalry between the two main fields of expertise in life insurance, life insurance medicine and actuarial theory, also accelerated the spread of standardized measures. First, the actuarial profession tried to replace the traditional individualized medical examinations by simple standardized calculations; and then, the insurance physicians reacted by developing their own medical form of standard assessment.

Finally, the case of life insurance points out the ambivalent effects of implementing standards for humans in institutional contexts. Originally, the category of substandard lives was designed as a threshold; being labeled a substandard risk was equivalent to being excluded from access to life insurance. But the definition of *substandard lives*, required for deciding who was to be excluded from being insured, made it possible to calculate more precisely the values of the different substandard risks at stake. Thus, gradually over the course of the twentieth century, more and more previously excluded illnesses were deemed calculable enough to grant the people who had them insurance under specific conditions (mainly a surcharge on the premium). By the 1960s, practically all previously excluded risks had become insurable. Indirectly and through a long process, the measurement of substandard risks led to the admission of more and more health risks to life insurance coverage.

There are three reasons for this inclusive trend from prohibitive to classificatory standards. First, insurance appealed to new groups of customers for economic reasons. As soon as statistical and medical research succeeded in calculating substandard risks, those groups of people became potential clients. Second, the professionalization of actuarial theory and life insurance medicine acted as a catalyst for insuring substandard risks. Both professions competed with their own assessment procedures in offering specific forms of contracts to previously neglected groups of the potential clientele. The question of how to insure clients with severe ailments eventually became a means to a different end—the institutional competition between two professions. Third, the insuring of substandard risks was also determined by insurance clients themselves. Their preference for general insurance companies with special offers for substandard cases, instead of specialized firms dealing only with substandard risks (because clients did not appreciate being stigmatized as customers of a substandard company), eventually led to the collapse of the specialized companies. In this sense, standards for humans are particularly flexible and self-reflexive because they partly depend on their own objects.

The Pernicious Accretion of Codes, Standards, and Dates

The People of the State of California v. Vernell Gillard

MARTHA LAMPLAND

Criminal law in the United States has long been an excellent site for demonstrating the ways that otherwise seemingly straightforward codes and standards can become distorted into frightening examples of irrationality. This has been especially true in California since the inception of the Three Strikes Law (California Penal Code § 667, subdivisions (b)–(i)). Instituted to identify and punish repeat offenders, the Three Strikes Law has become a great teaching tool illustrating how standardized punishment metrics can result in blatantly absurd sentencing consequences.

Yet, in itself, the punitive nature of Three Strikes is not unusual and falls easily within a familiar tradition in the history of U.S. justice. Less visible but equally powerful techniques for supervising citizens reside in mundane activities such as filling out forms, maintaining records, and having to submit to exams. What happens when willful deceit or perhaps negligence occurs when preparing documentation and then collides with intensely punitive sanctions? As the following brief review of *People v. Gillard* illustrates, there can be dire consequences for the offender. As a social problem, however, this case prompts us to ask what is the purpose of filling out forms, why do we maintain computerized databases, whose records are whose, and whose standards for punishment count? The answers to these questions may seem straightforward, but the following brief discussion of one law case demonstrates that these questions cannot be answered so simply or naively, or at least it forces us to question anew our assumptions about standards and forms.

- It is a felony to falsify or misrepresent one's history of worker's compensation claims in the state of California.

I am indebted to Carl Fabian, Esq., for providing me with documentation and valuable explanation of the case.

In the fall of 1993, Vernell Gillard had an accident at work. He subsequently filed a Workers' Compensation Claim. When he was sent to the insurer's doctor for a medical exam on October 13th, he (to quote from the 1995 verdict) "denied having previous back and right knee injuries." When Mr. Gillard initially sought medical help after his accident, he did report to the doctor then examining him that he had had a prior knee and back injury. His claim was accepted.

- A claims index is available to doctors and insurance agents that lists all previous Workers' Compensation Claims. This provides independent information that can be consulted at the time of the medical exam. If the claimant denies a previous injury (by deliberate deceit or from forgetfulness), the doctor can simply refer to the computerized database to challenge the claimant's position. This information is also readily available to the insurance company itself. It is also possible to conduct a court filing index search, which shows if any lawsuits had been pursued regarding physical injury or compensation.

- Workers' Compensation carriers subpoena medical records from the first examining physician prior to conducting their own medical exam of a claimant.

  When subpoenaed, the first doctor Mr. Gillard consulted did not send all of the medical records relevant to Mr. Gillard's previous injuries, including information that Mr. Gillard had acknowledged those at the time of his accident.

- Chapter 12 of the California Insurance Code § 1872.83 is entitled The Insurance Frauds Prevention Act. Provisions of this act establish a fund to be used to investigate and prosecute workers' compensation fraud (effective October 3, 1993). As subdivision (d) stipulates, 40 percent of the funds are to "enhance investigative efforts," while another 40 percent are to be distributed to district attorneys' offices "as to the most effective distribution of moneys for purposes of the investigation and prosecution of workers' compensation fraud cases and cases relating to the willful failure to secure the payment of workers' compensation."

  The discrepancy between Mr. Gillard's testimony to the insurance carrier's doctor on October 13th and his prior history, identified as fraudulent by both the doctor and the insurance agent based on the claims index, was brought to the attention of the district attorney's office. It bears emphasis that Mr. Gillard's ostensibly fraudulent testimony may not have become an issue beyond the confines of the insurance company and doctor's offices if the provisions of Chapter 12 had not made it profitable for district attorneys to seek out possible cases of insurance fraud they could prosecute.

- In December 1993, Mr. Gillard was deposed, at which time he did mention having had an auto accident in 1987 where he injured his "whole body" and was hospitalized for two weeks.

- In two separate medical exams—conducted on March 23 and April 8— Mr. Gillard denied "having a previous right knee injury."
- California Penal Code § 667, subdivisions (b)–(i) went into effect on March 7, 1994, approximately two weeks before the second series of medical exams conducted by the insurance company were held.
- On September 28, 1995, Vernell Gillard was convicted of three counts of insurance fraud, and one count of perjury. Because he had two prior felony convictions, Mr. Gillard was sentenced to twenty-five years to life. (The prosecuting attorney asked the judge to sentence him to 78–81 years to life.)

A case that had begun as insurance fraud, carrying a possible penalty of one year in county jail, or two to three to five years in state prison, had now been elevated to a case of Three Strikes. As the Notes to Decisions make clear, this case was appealed on several occasions. After his third appeal, the trial judge dismissed one of his two prior convictions and imposed a sentence of eleven years in state prison. He was released from prison on parole on November 16, 2003, after having served 8.5 years.

We can question Mr. Gillard's judgment or his memory in this case. He either willfully or negligently committed fraud, and he served time for that crime. He is not responsible, however, for the doctor's withholding information about his medical history and his acknowledgment of previous injuries when subpoenaed. The question arises: What is the status of information in databases, medical records, and legal depositions? While legal battles rage about whose truth is true, we are led to believe that the information contained in databases of legal suits and insurance claims are stable. Indeed, we are especially enjoined to believe that computerized databases assist in transparency, in stopping fraud before it starts, and denying inappropriate claims.

Yet, in this story, what standards of behavior are being demanded of Mr. Gillard and for what purpose? This is a single story of one person's travails within codes, databases, sentencing prescriptions, cash-strapped bureaucracies, and overeager industry watchdogs. Unfortunately, it is all too clear that this is only one of many, many cases that could be cited to demonstrate the pernicious accretion of codes, standards, and dates.

■ Standards without Infrastructure

ELIZABETH CULLEN DUNN

What are standards, anyway? Artifacts? Practices? A mode of governance? In *Seeing like a State*, James Scott (1988) argues that utopian "high modernist" techniques to improve the human condition are a form of *episteme*, or abstract knowledge. Certainly, bodies of standards fall into that category, too; as a set of codified rules, they exist in a placeless place where the complexities and differences of real landscapes and actual people are glossed over. When standards are used to dictate practice or to grade products, they often replace *metis*—the unwritten practical know-how that local producers gain over the years as they work to adapt to the ever-changing conditions of their lands, their markets, and their communities. Classifying standards as a form of knowledge has thus pointed attention toward the great power divide between standards-writers, who control the language of science and market opinion, and the local producers who are bound to obey these standards but who do not have the power to create them or change them.

But seeing standardization as a kind of epistemology—a practice that exists at the level of ideas—leaves aside the ways that standards interact with the material world and with material practices of production. If standards are about vaunting *episteme* and devaluing *metis*, they must also constantly engage with *oikodomi*—the infrastructures from which standards emerge and that shape the way they actually affect production. In many cases, having a certain sort of infrastructure—usually the same set of technologies that the people who originally wrote the standards had—is a prerequisite for entering the market in the first place. Rather than being governed by a set of complex rules that dictate how and what they will make and trade, producers without the right infrastructure are forced out of the market altogether.

In Poland, for example, meat producers have been confronted with a new set of sanitary and phytosanitary standards (SPS) that the country was required to adopt as a condition of EU accession. These standards lay out detailed specifications for slaughterhouses, from flooring and wall placement to the computer systems needed to implement farm-to-table tracking of each piece of meat. These systems may reduce the risk of food-borne illnesses such as trichinosis and bovine spongiform encephalopathy (BSE)—neither of which is endemic in Poland anyway—but they also have a significant side effect. Because the regulations are so stringent, they require most slaughterhouses to do significant renovations in their physical plant in order to comply. Most Polish slaughterhouses are small-scale village operations, however, and their owners do not have the capital needed to meet the standards.

The result has been that local Polish slaughterhouses have gone out of business in droves since the standards were adopted in 2003; one source estimates that more than two out of three abbatoirs and processing plants have gone out of business. Multinational firms such as Smithfield, the largest pork processor in the world, are poised to take over the market. This is likely not only to force more small-volume Polish-owned processors out of the market, but to force Polish farmers—most of whom have only a few pigs at a time—out of the market as well because Smithfield will rely on factory hog farms for a steady supply of pigs into their high-volume abbatoirs.

The point here is that, although harmonized regional standards such as those of the European Union are supposed to *reduce* technical barriers to trade, they often *create* technical barriers to trade when they enter locales that have markedly different infrastructures than those in the countries where they were developed. A standard without an appropriate infrastructure cannot be put into force without major upheavals in the physical environment and the social organization of production. Those who cannot bear the costs of these rapid transformations—like the Polish smallhold farmers—are excluded from the increasingly global market.

The problem of infrastructure is almost never raised in discussions of global standards, which are often assumed to operate in the ideal-typical homogeneous space of the world market. Countries are coerced into adopting standards, whether or not they are appropriate for local conditions of production. That is nowhere more true than in the former Soviet republic of Georgia, which is being required to adopt the standards contained in Codex Alimentarius. Codex (as it is known) is the joint production of the World Health Organization and the Food and Agriculture Organization. The rules that it prescribes for ensuring food safety and hygiene are endorsed by the World Trade Organization (WTO), which requires each of its member states to adopt Codex as a condition of entry. Georgia, as a new applicant, is busily trying to adopt Codex standards for industrial food production and to require all its food-processing factories to abide by them.

There is only one problem—Georgia is almost devoid of food-processing factories. Although the country had over seventy fruit- and vegetable-processing plants prior to 1989, the collapse of the Soviet Union and the resulting loss of the Russian market put all but five of them out of business. The producers that remain work on tiny budgets in primitive conditions. The second largest jam factory, for example, consists of three men, two rusted cooking vats, and two antiquated hand-operated pasteurizers in a run-down stone building.

How can entrepreneurs meet Codex standards in a society where the economy has collapsed, where the buildings are literally falling down and the electricity grid has not delivered reliable power in over a decade? Complying with Codex demands farm-to-table tracking and data collection that are best done with a computer and rigorous microbiological testing that requires a laboratory. These standards were written to reflect the so-called "best practices" in the United States and Western Europe, where the market is dominated by large processing plants equipped with gleaming stainless-steel tanks, swiftly flowing conveyor belts, computerized record-keeping, and in-house laboratories.

These two examples, Poland and Georgia, make two basic points about standards. First, although standards present themselves as *episteme*, as pure idea that exists outside of particular places, standards need an *oikodomi*, a material context in which they are transformed into action and effect. Fully analyzing standards requires seeing them within specific infrastructures and observing how they modify and are modified by particular physical environments.

The second point is that standards sort more than products. They sort producers into categories of worth, often based on their degree of difference from the powerful countries or actors, which write the standards to reflect their own infrastructures. Under this scheme, some producers can contort themselves and alter their infrastructures enough to be deemed fit; others can only partially comply and are deemed substandard. The producers and societies that are forgotten are the ones that cannot fit into the regulatory standard at all; because their infrastructures are so vastly different, they cannot be encompassed by the standards. Global standards thus may be aimed at increasing trade and opening markets to new producers, as advocates of the WTO or Codex proclaim. They may also be meant to be a means of "seeing like a state" and disciplining producers by making their inner workings more visible. In the process, however, they shunt some people and places to the margin and make them invisible to the world market. Those people and places should also be accounted for as we investigate the processes of standardization.

CLASSIFYING LABORERS

5

Instinct, Property, and the Psychology of Productivity in Hungary (1920–1956)

Martha Lampland

■ Twentieth-century class struggles took many forms: strikes, street fights, parliamentary debates, and new regulations. Battles were waged over the legitimacy of property and the rights of labor. At stake were the fundamental moral principles justifying social hierarchies. Coalitions on the left and right of the political spectrum proposed solutions to the discontent and to the economic turmoil it fostered. One crucial site for adjudicating the validity of social classifications was the field of work science. Promising social harmony, work scientists also held out the promise of greater efficiency. Armed with innovative techniques to improve performance, work scientists set out to rationalize the physical actions of laborers. Equally important was the discovery of workers' innate qualities and psychological states. At the heart of these activities was the search for mechanisms: the mechanics of action and the motivations for behavior. Once identified and understood, these mechanisms could be marshaled to increase productivity. The search for successful strategies of rationalizing the workplace entailed experimental work, extensive training to solidify the standards of performance embodied in the innovations in task performance, and, not least, occupational testing to find the right person for the job. And it is this productive tension—among scientific engineering, standardizing, and social

I acknowledge the valuable assistance many colleagues provided me while I wrote this piece. In addition to thanking my colleagues in the research group at Irvine, I also received valuable comments from Thom Chivens, Elizabeth Dunn, Jim Fleming, James Mark, Alex Stern, and two anonymous reviewers. The work leading to this article was supported in part from funds provided by the National Council for Eurasian and East European Research, for which I am grateful. The Council, however, is not responsible for the contents or findings of this report. I also thank the International Research Exchanges Board, the American Council of Learned Societies, and the German Marshall Fund for their support of my research.

classificatory practices (racializing and classifying)—that constitutes the focal point of my analysis explicating the powerful scientific and political techniques used in creating people and justifying hierarchy.

In this account, I review the strange history of agrarian work science in mid-twentieth-century Hungary, in which, as we learn, a schema accounting for the mechanics of work in capitalism was sustained in socialism while the moral valence of its classificatory principles were inverted. In both the capitalist period of interwar Hungary and the Stalinist era of the 1950s, people's motivations were understood to be given, habituated, and even instinctual, but, curiously enough, instincts varied according to an individual's ethnic background, property relations, and family history. The capacity to work was not simply learned; it was fully embodied. Yet despite the presumption that classes and ethnic groups carried specific work habits in their physical makeup, the attribution of diligence and sloth were reversed in the course of a few decades. So what would otherwise have been considered very different social strata—landless workers of the 1930s and wealthy landowners of the 1950s—came to be classified into the same category. Both were, by their very nature, incapable of working. The dangerous subversive of the interwar era—the unemployed worker *(lumpenproletariat)*—was considered to be as subversive in his time as the wealthy peasant *(kulak)*, the prototypical class enemy in Stalinist villages, was a decade later. These sentiments were voiced in political speeches and pamphlets, although the sources I cite here are more obscure: treatises on rationalization and agrarian work science texts from the 1920s and 1930s, and bureaucratic reports on Stalinist cooperative farms from the early 1950s. Standardizing occupational categories relied on the scientific development of theories of motivation and human behavior, categories that built explicitly on, and further elaborated, local classifications structuring social inequality: class, race, and gender. Social inequality was solidified, indeed authorized, by modern projects of standardization, not despite of but because of the ways that scientific practices were crafted.

In common parlance, to standardize means to introduce uniformity and/or to align with a stipulated goal. Standardizing workers may suggest imposing a homogeneous template, echoes of which are found in classic critiques of the assembly line and deskilling (Braverman 1974). This cookie-cutter view misrepresents or underestimates the degree to which standardization entails complex techniques of differentiation (see Steve Epstein, chap. 2; Martin Lengwiler, chap. 4; and Judith Treas, chap. 3 in this volume). The standardization of workers central to scientific management strategies never totally erased differences between workers, just as it never eliminated the division of labor itself. Pernicious differences that were seen to impede production were squashed, but other differences were sustained and even enhanced by the techniques of work science. Complementary to these efforts was the focus on developing uniform scientific techniques for differentiating workers, for molding groups to particular specifications. Once perfected, these practices would be offered in the marketplace of ideas for modernizing society. Work scientists and their colleagues in scientific management and occupational psychology envisioned this to be a worldwide market because the techniques being designed were formulated in terms of instinct and human nature, shared by all. Yet the actual practices of creating a science and managing subjectivity in the workplace entailed codifying cultural-historical assumptions shared locally by scientific practitioners about the subjects (and subjectivity) of analysis. In short, sociocultural norms justifying existing social hierarchy were refined into scientifically tested principles of modern organizational forms. Science and industry clasped hands, committed to the grand project of classifying work and standardizing workers.

It is clearly not earthshaking to suggest that twentieth-century scientific management principles reinforced existing patterns of inequality and exclusion; the question, of course, is: Why? More to the point, what are the hows that explain the why? We could easily list a series of conventional explanations, from crude Marxist explanations painting pictures of mean-spirited executives reveling in exploiting the lowly worker to far more sophisticated studies of social capital and the reproduction of inequality (Bourdieu 1984, 1986). My intervention into the debate comes from a different angle—grounding the study of science and engineering in social histories of subjectivity and political economy. I argue that theories of human nature, and indeed instinct, carried very local meanings; they were socially inflected in important and divergent ways, contrary to what we might otherwise assume about such universalizing categories. Everyone agreed that morality and motivation were embodied, but the social location of virtue differed, depending on the social position of the observer. This becomes clear in the history told here—the principles justifying social inequalities changed radically, but the mechanisms explaining motivation, as well as the belief that capacities were physically embodied—remained the same. So, it behooves us to consider the social context within which these ideas were developed and deployed, that is, how, in the name of social engineering, standardization strengthens existing explanations for human difference while altering the means by which these differences matter. In other words, in the following analysis I offer a more sustained investigation of the social as constitutive of work science, in particular, and of social science, in general.

This story may be arcane, carrying all the negative connotations of mid-twentieth-century scientific engineering, eugenics, and political indoctrination.[1] I hasten to add, however, that the language of human nature and biological explanations for behavior are just as common in academic and business studies today—economics, psychology, management, and political science—as they were fifty years ago. This short exploration is intended, therefore, to give us reason to question

1. Social engineering tends to be associated with what are now considered morally reprehensible practices, such as eugenics and Stalinism, but this substantially misunderstands the centrality of transforming individuals into efficient workers and voracious consumers in the dynamics of capitalist economies.

the meanings of human nature and the purposes of genes in current social analyses, as much as to ponder the consequences of theories of instinct and politics in the historical example at hand.

Studying the Soul, Rationalizing the Job

Agrarian work science is perhaps the least known of the cluster of fields subsumed under the name scientific management, which flourished in the mid-twentieth century. Scientific management writ large encompassed the work of many disciplines: psychology, economics, physiology, business, accounting, and engineering. This was just as true of agrarian work science as of studies of industrial production. In Hungary, for example, we find studies in agricultural magazines and specialist journals concerning nutritional requirements and energy expenditures for specific tasks; time and motion studies to enhance workplace performance; and experiments with organizational patterns, shop floor design, and wage structures to increase productivity (Bereznai 1943; Blantz 1936; Fischer 1936; Kemptner 1939; Keszler 1941; Kölber 1932; Molnár and Schiller 1937; Schiller 1940; Schrikker 1942). Careful measurements of output and experiments with tool use were rigorously executed, and results were displayed in graphic tabulations and mathematical formulae. Studies of the movement and rhythm of agricultural work, such as cutting hay (Szakáll 1943; Ujlaki 1943), assisted agrarian work scientists in calculating the amount of time required to perform this task. Although clearly influenced by industrial work science, agrarian work science traced its lineage in Europe to manorial estate management as much as to economic studies of the firm.[2] This did not lessen its commitment to calculation and measurement. Ludwig Ries, an influential German specialist, encouraged estate managers to carry a slide rule at all times to perform the calculations necessary for time studies; he confessed to feeling more uncomfortable on the job without his slide rule than without his watch (Reichenbach 1930, 234). Béla Reichenbach, author of the definitive study of agrarian work science in interwar Hungary, listed the following aids to the study of agricultural work: stopwatch, slide rule, dynamometer, psychographs, respiratory devices, and moving film (222).

Studies of the physical conditions of laboring were closely tied to the study of workers' dispositions, constituting the fields of psychotechnique, industrial and occupational psychology.[3] Occupational psychologists or practitioners of psychotechnique constituted only one branch of the relatively young and exciting

2. I am indebted to one of the reviewers for calling my attention to references to agrarian work science in the United States (see, e.g., Friedland, Barton, and Thomas 1981). A history of manorial estate management and a focus on labor were not the primary focus because mechanization was at issue.

3. "The science of work is divided into four branches, in accord with the four-fold nature of work (intellectual, psychological, physical and social): 1. the philosophy of work, 2. the psychology of work, 3. the biology of work, and 4. the sociology (organizational studies) of work" (Rézler 1944, 13).

field of psychology. As concerned as their more famous colleagues in psychoanalysis with the fate of the human condition, they sought to realize their goals of bettering humankind on the shop floor rather than on the couch. Their prosaic task did not exclude them from participating in the heated debates raging throughout the discipline at the time (and since) over the relative significance of biology and culture, of instinct and behavior (Degler 1991; Rabinbach 1990; Stern 2003). Unlike colleagues whose work issued from mental institutions or laboratories, however, their immediate inspiration came from studies demonstrating the wide variation in abilities to perform particular tasks or master specific skills conducted by specialists committed to rationalization, scientific management, and work science (Hatvani 1935, 3).[4]

Work science labored to render workers' abilities visible, calculable, and hence malleable (Rose 1990). Analysts wished to ascertain the psychological features— skills and temperament—needed in specific jobs and then to develop tests to identify these qualities in job applicants. A number of scientific management studies, such as the Hawthorne experiment, had attempted to establish the influence of various environmental and emotional variables in work.[5] Comparable efforts were devoted to fine-tuning wage schemes. The design of new wage systems to improve productivity had to consider the relative ability or inclination of workers to perform effectively on the job by matching ability to activity and temperament to context. Some qualities were considered more amenable to change and others less so. In light of these constraints, it may have been difficult to change people's behavior—to find just the right sorts of incentives and exercise adequate pressures to influence actions—but it was for this reason all the more worthwhile. The pragmatic tasks of increasing productivity in the depths of the worldwide depression fueled excitement about scientific management, work science, and occupational psychology. Aspirations were loftier, however, because they were nurtured by utopian dreams of a perfectly functioning, well-oiled social machine, unmarred by distrust or unrest.

A curious tension is embedded in the nexus of psychology and social engineering in work science. The purview of scientific discovery and the role of social design are intermeshed and difficult to untangle. Analysts committed to understanding workers' aptitudes and temperament—economists, as well as psychologists—regularly invoked the language of human nature, thereby positing a universal set of motivational

---

4. Koyama Eizos' projects of ethnic engineering and public opinion surveys in mid-century Japan show strong similarities to work science in Hungary in the 1930s and 1940s (Morris-Suzuki 2000).

5. Studies of environmental factors included investigations into the influence of climate on productivity. Ellsworth Huntington investigated the relationship between weather and efficiency in the 1920s when he occupied the chair of the Committee on the Atmosphere and Man of the National Research Council of the U.S. National Academy of Sciences. In his study of climate research, James Fleming refers to Huntington's research as "meterological Taylorism" (1998, 100–103).

traits. Instinct was also used to explain patterns of laboring, connoting a transhistorical and acultural human physiology. Engineers designing new machines or business managers introducing a new wage system recognized the constraints of natural impulses, yet much of their work revolved around retraining workers using scientifically calculated strategies to eliminate behavioral traits that impeded their schemes for improved efficiency.[6] Yet how do we distinguish between proper motivations and problematic behaviors, if in fact the reasons for action are understood to be natural or instinctual? Just how do we reprogram human nature or erase instinct? This was precisely the difficult task that occupational psychologists and work scientists grappled with in their studies of, and attempts to alter, work and workers. They confronted the challenge eagerly, marrying their scientific interests to the pressing need for economic renewal.

Designing a Modern Hungary

Following World War I, Hungary was in a political and economic shambles. Hungarian politics was cut loose from the crumbling edifice of the Austro-Hungarian Empire and, in a few short months, lurched from a liberal republic to a people's (Soviet) republic. Conservative forces rallied, dismantling soviet institutions, such as town and worker councils, and murdering suspected communist sympathizers. The economy was in ruins, battered by prolonged fighting. More important, however, was the loss of two-thirds of the nation's territories to the new successor states in Eastern Europe, lands once integral to the national economy. The loss of valuable resources, as well as numerous customers, crippled the nation's economy at precisely the moment at which the fortunes of many nations were deteriorating as well. Once the backbone of the national economy, agriculture now became its emaciated skeleton. Feeble attempts at land reform did not challenge the dominance of manorial production. Fifty-six percent of arable land remained in the hands of the aristocracy, Catholic Church, and wealthy peasantry; the remaining 44 percent was worked by small landowners (Weis 1930, 162). Two-thirds of the agrarian population—one-half of which was officially defined as landless, the other half too poor to sustain their families—were forced to cobble together a living as best they could (Kerék 1939, 348). Some tried to provide for their families by working a few acres, but the majority could not subsist on their paltry holdings and ended up as residents on large manorial estates or as impoverished daylaborers in villages and towns.

The sorry state of agricultural production, even on huge farms owned by the wealthy elite, hindered the growth of the national economy. The equitable distribution of land was held out by many as a solution to economic stagnation, but

---

6. It bears mentioning that psychological studies were not restricted to workers but were extended to the psychology of management as well.

another group—comprising business economists and work scientists—advocated modernizing production. They hoped to follow the lead that their German colleagues had shown in developing scientific management techniques for agriculture. (They were conversant with similar innovations in the United States, Italy, and the Soviet Union as well.[7]) Agrarian work science promised to achieve higher profits and lower costs, which were seen as the most effective (and perhaps sole) means of competing in the dismal international market. In the long term, however, their goal was to modernize Hungarian agriculture by reforming its organizational structures, improving economies of scale, mechanizing production, introducing accounting standards, and transforming small property owners into a more efficient labor force. Their activities during the interwar period consisted of establishing departments of agrarian business economics in several institutions of higher learning, training students, writing columns in influential agrarian newspapers, and participating in research activities funded primarily by the Ministry of Agriculture. It is to the products of those efforts that I now turn to examine the manner in which theories of behavior and notions of human nature were articulated.

Manorial estate managers and work scientists shared the views of their Hungarian compatriots nationwide that workers differed in their skills and abilities, according to their regional identification, racial/ethnic identity, class position, and gender (e.g., see Turda 2007; Turda and Weindling 2007). The ethnic/racial classificatory schema that work scientists used had deep roots in parochial and broader territorial squabbles. Long histories of animosity between adjacent villages or counties bred prejudices among fellow citizens, expressed frequently in terms of working habits and moral behavior. Similar prejudices were elaborated in the nationalist battles over economic authority and political sovereignty in the decades leading to the demise of the Austro-Hungarian Empire, attitudes that lived long after the Hapsburgs had left the throne.

There are substantial differences between agricultural workers ethnically. With regard to working capacity and suitability as a manorial servant, day laborer or contractual laborer, the pure Hungarian is worth the most. Among these, the people from the Great Plain and Dunántúl are especially so. Even though we pay these workers well, they nonetheless work the cheapest because of their substantial performance. Following them are the Slovak workers from these regions, where they haven't become too americanized[8] [lazy]. Their working capacity, however, is less.

7. In 1920, the Ministry of Agriculture in Hungary established the National Agricultural Business Institute (Országos Mezőgazdasági Üzemi Intézet), modeled on comparable national institutes in Austria, Denmark, Germany, Italy, Norway, and the United States (Károly 1925, 2–3).

8. The term *become americanized* in Hungary means to become a reluctant or recalcitrant worker (*Révai Nagy Lexikona* 1911). This is based on the active resistance American workers made to the introduction of Taylorism and scientific management, a history well enough known in Hungary at the time for this to become a commonly used expression.

German workers are excellent, if they are working for themselves, but they don't gladly engage in wage labor, especially for a master of another ethnicity. Moreover, they are expensive; together with the Serb they are the most expensive worker. The undemanding Ruthenian and Romanian are fairly obedient, but they are very weak workers. Although they make do with less, they perform very expensive and slipshod work. (Reichenbach 1930, 247–48)

Reichenbach's description of ethnic workers in agriculture living within the country's borders could have been drawn from any number of social science texts in interwar Hungary. The disciplines of Hungarian ethnography and rural sociology were firmly grounded in identifying and explaining differences among classes, races, ethnic groups, and religious communities within the nation (e.g., Erdei 1941; Győrffy 1942; Illyés 1936; Kovács 1937; Szabó 1937). The experimental and observational work of occupational psychologists and work scientists confirmed these views.

[O]ne must lead a people inclined to a life in the barracks entirely differently than the Hungarian worker, in whom there is a strong propensity for independence and a tendency to take on responsibility. [Hungarians] have a difficult time enduring rigid military training. Therefore one must establish work psychology separately for each country and each people. Without preliminary studies from abroad, we must undertake this significant duty ourselves. (Rézler 1944, 14; see also Thurzó 1934)

Distinguishing the dynamics of different social groupings also entailed founding several branches of psychological investigation: folk psychology, social psychology, and mass psychology (*néplélektan, társas-lélektan,* and *tömeglélektan*; Rézler 1944). Scientific rigor demanded the creation of subfields to investigate the specific forces at work in varying social configurations, providing insights into the causes of difference and the means of domesticating variety in the service of standardization.

Hungarians were not alone in distinguishing workers by their ethnic or national identity. The premier German specialist in agrarian business economics, Friedrich Aereboe, articulated similar sentiments in a lecture he gave in 1930 describing his travels in the United States the year before. Contesting the usual explanation among Europeans for U.S. economic domination, which focused on natural resources and the low costs of acquiring land, Aereboe argued that the human factor was the most significant difference. The United States was settled primarily by "the northern races," who enjoyed superior "mental and physical abilities" (1930, 6). Recognizing the poor educational level of U.S. workers, Aereboe countered that racial characteristics compensated for this deficiency, especially if one surveyed the specialization of nationalities in agricultural production:

The cooperation of the different tribes [*Volksstämme*] with their distributed roles has advanced America immensely. The Germans soon proved themselves to be the best crop farmers, the Scandinavians and the Swiss the best in animal husbandry, the talented Celts from the green island as the best speakers, agitators and traders, the Italians as the good innkeepers and fruit traders, while the English with their great organization and leadership talents quickly took on everywhere a comfortable leadership role. (6)[9]

In light of subsequent events in German politics, Aereboe's ethnic division of labor seems tame. Radical notions that skills were embodied in racial stock can also be found among Hungarians, for example, in Jenő Kodar's (1944) discussion of race psychology.

Just as common in these texts are differentiations of ability according to class position. The social hierarchy of rural communities between the wars was built on the rock of property ownership, which became the guarantor not just of an individual's livelihood but also of his or her moral character. Having to leave home to find work, that is, being relatively mobile in the labor market, was a sign of a lack of control and so a loss of integrity (see Lampland 1995). We see these sentiments clearly expressed in Reichenbach's differentiation among workers according to property ownership:

A portion of day laborers are dwarfholders [i.e., own very small properties]. They are more class-conscious and so in general more difficult to manage, but they are more diligent, reliable and skilled workers. The day laborer without property either lives in his own small house, or usually is just a tenant; he often doesn't even have his own domestic animals. The latter are the most mobile, and the least reliable workers politically and morally. (1930, 196)

Poorer workers were not entirely unreliable. A family history of faithful service at the manor—a respectable form of diligence—bode well. "Good traits are inherited over generations. That is why we attempt to obtain as many decent, capable and diligent families as possible working permanently for our estate. At our large estates we often find workers' families, which are descended from many generations of excellent manorial servants and workers" (Reichenbach 1930, 247). On the other hand, not being primarily employed in agriculture robbed one of the ability to work effectively, as the following passage indicates. "Among migrant workers one may also find village craftsmen and industrial workers, who take on migrant labor in the summer to ensure a better living. These latter are usually the least fit for use in the work gang, so it is best if we don't allow industrial workers to be accepted into the crew" (209).

Where an individual worked and how was determined by his or her social history, but this was intimately related to his or her psychological state. In an article written by a manorial steward who had conducted experiments in innovative wage schemes, we learn that there are three pillars to rationalizing work. The first two are the introduction of labor-saving devices and rationalizing the division of labor

9. I am indebted to Shannon McMullen for helping me wade through this text and for providing translations.

and the practice of management. The third is taking advantage of "psychological elements," that is, making the worker interested in the result of work. As he explains, "This has great significance, because the farm-owner profits by this system since the worker produces greater value and the worker also shares in the higher proceeds" (Kemptner 1939, 33–34). We see, therefore, that stratifications of laboring skills according to ethnicity and class were paired with the instinct for material reward. Deep in the heart of all workers was the "natural" desire for economic gain, as Rezső Károly, the Hungarian proponent of agricultural Taylorism, explains, "The greatest disadvantage of time wages . . . is that the worker is under no compulsion, since he has gotten used to the overseer or attendant at his heels. This compulsion is never as effective as the stimulus which material gain represents" (1924, 78–79). This stimulus was powerful enough to reverse unfortunate historical trends. "The World War and the revolutions that followed corrupted workers The sense of duty and conscientiousness have diminished. Therefore one must exploit the base and simplest instinct of workers—greed—for the effective increase of work" (Reichenbach 1930, 239).

A variety of terms are used in Hungarian texts to refer to the psychology of working. In one text, an advocate of rationalization cites a mood or inclination for work (*munkakedv*). "Selecting proper tools and proper human material, as well as establishing proper work norms have the purpose of regulating personal questions of rationalization, yet perhaps of these the most important is the question of work mood" (Raith 1930, 256). Károly (1924, 77, 80) mentions inclination and instinct, whereas Reichenbach uses the term inclination but also discusses workers' nature, their wishes, and their interest in diligence or, more generally, psychological motives as well as creative instinct (*alkotó ösztön*) (Reichenbach 1930, 200, 225, 228, 237, 265). Various terms—instinct, motivation, inclination, and interest—are used interchangeably in these texts. This was not unusual at this time because psychologists worldwide had not reached consensus on their use. The relative significance of various factors in human behavior—evolution, heredity, genetics, culture, and personality—were at the center of debate within psychology and related disciplines. For example, in the United States, instinct theory had faded from psychology by the 1920s, but drive theory had taken its place (Cravens 1978). In the very different world of psychoanalysis, the concept of instinct suffered from being underspecified analytically. I quote here from a Hungarian in the psychoanalytic tradition: "There are primary repressed energies in the psyche, which various psychological schools call instincts, needs, wants, etc. All that we can really determine is that repressed energies exert an influence at the depth of the psyche" (Révay 1946, 86).

All these terms connote innate or embodied qualities crucial to the laboring process, conflating theories of social difference and natural desires for material reward. In principle, we might assume that these qualities could coexist without having to be implicated in one another. The reason I do so is that they are tightly intertwined in the cultural history I am analyzing. Hungarian theories of behavior partook of both these notions: socially constituted features and biologically determined responses.[10] Socially differentiated features were embodied and far more durable than the prevailing view today in the social sciences that human behavior is learned through social interaction. Embodied differences were further grounded in assumptions about universal characteristics of human nature, such as the enduring desire for gain.[11] It was the task of work scientists to skillfully manipulate these variables of the work process to economic advantage. "In our interventions in the interest of increasing the yields and income of peasant farms, we must give our full attention to the mentality and emotional experience of the common man, which has taken root in and been imprinted on the peasant brain and soul from generation to generation" (Kulin 1933, 57). In other words, mentality and experience were not ephemeral phenomena, but were converted into durable features of a family, ethnic group, or class segment. Certainly this is a more Lamarckian conception of the inheritance of traits, but it is no less a theory of inheritance for that.[12]

The variability among workers was a widely recognized social fact, one that presented a serious problem to modernizing and rationalizing the workplace. Indeed, the nature of workers was considered as substantial an impediment as the weather. "In connection with motion and tool studies we must emphasize the great difficulties which these studies confront in agriculture. These are caused by changing weather conditions, the nature of the workers, not to mention the extraordinary diversity of agricultural tasks. Workers . . . cling to custom too strongly; it is difficult to accustom them to different movements or different tools" (Reichenbach 1930, 228). Reichenbach goes so far as to question any customary method or style of work, following the advice of a famous German specialist in agrarian work studies. "According to Seedorf, we must consider every pre-existing work style which follows custom to be bad, and must search for a better one in its place" (232).

All the differences in play—class, gender, local custom, and ethnicity/race—were considered crucial variables for schemes improving productivity. Hence, we see that, although rationalizing the workplace entailed techniques to standardize the labor

10. The word for instinct in Hungarian (*ösztön*) means to motivate, stimulate, or urge (*ösztönöz, ösztönzés, ösztönző*) (Bárczi 1941, 230–31). This makes the original meaning of the term *prodding stick (*circa 1600), which shares the same etymology as *instinct*, more evident in Hungarian than in English. The notion of an embodied feature or characteristic came later, as was also true in English.

11. It warrants emphasis that "an enduring desire for gain" is assumed to be universal, but, in fact, it is not. As the historical and ethnographic record demonstrates, the desire for gain is not universally shared but, instead, is the product of very specific cultural and historical processes. See Lampland (n.d.b) for a discussion of the strong resistance that wealthy landowners in Hungary mounted against using the promise of monetary reward to entice villagers to work in the immediate postfeudal period (circa 1850s–1870s).

12. It is important to remember that the so-called evolutionary synthesis, that is, the reconciliation of Mendelian genetics and Darwinian evolutionary theory, was only in its infancy in the 1930s (Mayr and Provine 1980; see Smocovitis 1996 for a critique of the conventional history).

process, there were limits to the standardization of laborers. "Big differences exist within a group between the ability and willingness of individuals to work, even if the group appears to be homogeneous. The industrious, diligent and skillful worker cannot express his full abilities, because he is prevented from doing so by the weaker, less diligent, or just bad tempered workers, who cannot even be impelled by a premium to take greater advantage of their physical strength" (Fodor 1925, 44). Obstacles did not have to be conscious; resistance could also be conceived of as a simple engineering problem, formulated in the terms of constitutional or organizational resistance (Gárdonyi 1933).[13] Most texts, however, showed a more subtle understanding of the behavioral features of the human organism, which could be engineered, certainly, but not in quite the same fashion as machines.

So, if instinct and human nature are impediments, what can one do? One mobilizes all resources to transform the character of workers, to teach them to respond differently, and to master new skills. Work scientists and occupational psychologists devised training and skills testing for workers to complement changes made to the workplace itself. They were well aware, nonetheless, that the transformation of the workers would take time. As Károly reminded his readers, the full realization of Taylorism would be achieved ideally once the workers themselves accepted and internalized its principles.

> It would be a mistake to believe that the increased work results achieved by the Taylor system comes from greater brawn and intellectual exertion. In fact the result derives from the correct organization of work, the invention of the most expedient labor process, from saving wasted time, and the better use of machines and tools. *The greatest moderation must be exercised with the introduction of the Taylor system.* Years are needed to adopt it completely. One shouldn't force the introduction of the system. Rather one must attempt to have the workers appreciate the advantages of the system and be willing to adopt it themselves. (Károly 1924, 80; italics in the original)

In short, workers could be taught to recognize the advantages offered by scientific management and would come to embrace it on their own. More precisely, they would (and could) be taught to respond instinctively to new kinds of incentives and changes in working environments. The workplace would be peopled with units standardized to a range of labor specifications.

The job of raising the people's consciousness and moral character was also to be shouldered by school teachers and public intellectuals. Moral improvement—cultivating the folk (*népnevelés*)—would ensure the optimal improvement of the entire society. "The dexterity, diligence and reliability of time wage agrarian workers are not only increased by being bound to the land and a little property,

---

13. Gárdonyi's use of analogies from the physics of energy and discussions of organizational fluidity reminds us of the crucial role of thermodynamics in the early history of work science and even of political economy, for example, the work of Karl Marx (Rabinbach 1990).

but also by the intellectual and moral level of workers, so this is why people's education is such a truly important general national interest" (Reichenbach 1930, 196). Finally, however, one had to rely on the state, which had already shown itself to be effective in designing techniques of rationalization. The military was an important and early site for experimenting with and implementing standardized techniques in the United States and Germany (Kevles 1968; Rabinbach 1990; Rose 1990). This reminds us that the range of scientific research and education was not confined to productive sites. And it explains why work scientists continued to look to state institutions to assist in this broad effort.

> More and more the modern state aspires to an ideal condition, where it does not trust changing social and economic relations to the incalculable whimsy of fate. . . . To solve the new tasks of the modern state (*the one best way*) most feasibly, scientific institutes are established to research and determine the necessary concrete knowledge in various problem areas. In doing so they substantially contribute to an active state leadership resolving questions in an informed, professional and quick manner. (Rézler 1944, 16; italicized phrase quoted in English in original)

Advocating an interventionist state and (more) planned economy sat comfortably in this world of innate abilities and physical constraints on change.[14] It continued to be the preferred role for government in the postwar era, and particularly so once the Communist Party took control of the state in a few short years.

Socialist Wages and the Psychology of Collectivization

Agrarian business economists moved in influential circles during the interwar period. They had access to valuable resources in educational institutions and research, and published in prominent journals and newspapers. Nonetheless, they had little to no impact on the way agriculture was organized and laborers remunerated.[15] Their failures could be attributed in large part to the dismal condition of the economy. Wealthier farmers were unwilling to expend much effort on experimenting with their businesses in so precarious an economy and were certainly reluctant to replace payment in kind with monetary salaries, a central pillar of work science proposals. The possibility that agrarian modernizers could influ-

14. Planned economies and interventionist states were common in the interwar period (e.g., Germany, Italy, Japan, and the United States). It bears emphasizing that socialist state planning in Hungary grew out of an already very interventionist state climate bequeathed from the war years and postwar reconstruction (see Lampland n.d.b).

15. Mary Nolan's fascinating 1994 book on debates about Fordism and rationalization in Germany in the decade following the First World War shows a comparable disjuncture between intense interest in scientific modernization and very little concrete success in implementing these strategies at the level of the factory.

ence Hungarian production was threatened further by the land reform of 1945. Nearly all manorial estates over 200 acres were distributed to the poorest of the poor, smashing the economy-of-scale that agrarian business that economists considered foundational for modernization. This bleak period did not last long, however.

The Communist Party advocated the collectivization of agriculture within months of taking full control of the government in 1948. Collectivization was a crucial element of industrial modernization, intended to reduce the size of the agrarian labor force and ensure the availability of food to nourish the new proletarian paradise. It was also part of a broader commitment to reorganizing the economy along the scientific principles of Marxist-Leninist political economy. Agrarian business economists thus found themselves in the curious situation of advising the new socialist state on the development of wage schemes, despite their pedigrees as capitalist technicians. After all, crucial features of the new socialist science were shared by both socialist and capitalist camps, specifically the techniques of work science and rationalization. The same theories of instinct continued to inform Stalinist practices, but the social classificatory system and moral world within which they were deployed was inversely related to its capitalist forebearer—now the former landowners were to be reviled and the poor worker praised. Ethnic differentiations did not disappear from the classificatory schema, but they were secondary to the relentless focus of the Communist Party on class relations.

Labor relations in socialist industry as well as agriculture were built on the simple principle that higher incomes motivate greater productivity. The effective use of pecuniary motivation was the foundation for the structure of wage contracts at socialist enterprises and cooperative farms. To provide even greater impetus, additional features such as bonuses and premiums were developed, as well as the famous labor competitions held in anticipation of every possible communist holiday, party leader's birthday, and historical commemoration.[16] Unfortunately, agricultural production at cooperative and state farms did not increase; throughout the early 1950s, private farmers were always more productive. This is not surprising, despite the state's promotion of cooperative agriculture. Cooperative farms were poorly equipped; they lacked adequate tools and were manned primarily by formerly landless workers whose skills in running a farm were limited by their past work experiences. Large numbers of cooperative farm members simply refused to work the land owned jointly by farm members, preferring to work the small parcel of land they kept as part of their membership contract. Moreover, the membership of the farms was constantly changing. New members joined to escape the oppressive state tactics mounted against peasants. When the policies of the state became less punitive, families left en masse, leaving their fellow farm members in the lurch. State farms were

16. Stalinist competitions are usually depicted as primarily forcing workers to work harder, typified by the Soviet Stakhanov in the 1930s. Actually, competitions were held for a variety of goals: to reduce waste, encourage innovation, improve bookkeeping, and rationalize administrative tasks (e.g., see Gille 2007).

more successful, better equipped, and less subject to the vacillations in membership that endangered yearly productivity at cooperative farms.

The state worked hard to cripple wealthier peasants, primarily with extremely high taxation rates. The Communist Party attempted to restrict the private hiring of laborers, ostensibly in the interests of eliminating capitalist practices within agricultural production. The party/state also forbade private farmers from using any sort of payment by result, clearly wishing to keep the power of material incentives under its own control.[17] Giving managers incentives for individual performance was necessary, in concert with worker incentives, to ensure all efforts would be expended to increasing productivity.[18]

> In place of the current complicated, bureaucratic premiums that are not based on individual performance, the Council of Ministers wishes to introduce a system—based on individual performance suited to the principle of a socialist wage system—in order to reward the upper and middle management overseeing production (directors, department heads, group heads, etc.). The purpose of the new system is to make the upper and mid-level directors of production more interested in the increase of production.[19]

Bonuses and other tools to motivate the laboring community were an integral part of socialist wage schemes. Yet these flourishes on the basic incentive wage system of cooperative farms did not achieve their desired effect in rural communities during this period. The process of implementing new wage systems was marred by numerous problems, not least of which was the profound mismatch between party/state policy and local conceptions of a just reward for labor performed (see Lampland n.d.a).[20] These difficulties were not lost on party officials, as a document from a meeting of the Political Committee of the Hungarian Workers' Party in 1954 shows:

> The development of effective large scale production at cooperative farms has been prevented by the Ministry of Agriculture and other state organs not keeping the increase of cooperative farm members' monetary interests in mind. . . . The members' interest was reduced substantially by the exaggerated restriction of financial management, free market sales, and distribution of advance payments at cooperative farms, as well as the coercive character of state production plans. . . .[21]

17. "Performance-based work in the private sector must be prohibited" (MDP 276 f., 53 cs., 23 ö.e., 1949, March 17).

18. Stalinist society was not characterized by an oppressive collectivism. This image of collectivist culture is inaccurate, fostered by kitschy Stalinist newsreels as well as the many critics of Stalinism. Collectivism characterized public ritual, no doubt, but it finds little validation in the world of factory production or cooperative farm work. There a focus on individual effort, reward, and punishment was absolutely crucial to ratcheting up labor productivity.

19. FMTÜK, XIX-k-1-c 5. d., 1950, April 28.

20. Industrial workers, more familiar with payment by output, were far more belligerent and resistant to these innovations, simply refusing to work or physically assaulting local party figures when norms were revised downward (see Pittaway 1999).

21. MDP 276 f., 53. fcs., 156 ö.e., 1954, January 13.

The party/state hurried to remedy its failed policies, but by the fall of 1956, it had lost control over a range of public institutions, as open opposition became more frequent. After the Soviet invasion crushing the revolution and reinstating the party/state, collectivization was put on hold indefinitely. Only under significant pressure from the Soviets in late 1958 was collectivization started up once more, although this time the intense pressure and frequent use of violence against village communities resulted in virtually all private farmers joining cooperative and state farms for good.

We could argue that this concern with material incentives was merely a short-term concession to the capitalist mentality of the labor force. Soviet society and its Eastern European offshoots had committed themselves to the utopian project of creating Socialist Man and ushering in communism. The party/state knew far too well, however, that this radical change would not happen overnight, or even within decades.[22] Accordingly, communist officials had to work within the constraints of human nature. As the party secretary in Zala County reminded his subordinates, "We absolutely must reckon with the psychology of our peasantry."[23] In short, Hungarian socialism was not committed to social equality. Following in the tracks of Soviet ideologues and heeding Stalin's campaign against "bourgeois egalitarianism," Hungarian Stalinists promoted social justice on the principle that those who work shall prosper, in direct relation to the quantity and quality of their work. It was assumed that many—particularly the irredeemable "class alien"—would not be able to engage in honorable labor, thus "earning" their impoverishment. For this reason, it was absolutely imperative to purge socialist enterprises of class enemies whose nature would rebel against socialist morality.

> All across the county the enemy is attempting to worm its way into cooperative farms. This year we have already expelled 24 kulaks and other hostile elements, but there are still 7 cooperative farms, where there apparently are hostile elements, which will be expelled within a short time. The enemy's assault is now being manifested in their stealing their way into our cooperatives which they then destroy from within.[24]

New socialist institutions had to be protected from class enemies' impulse to demolish them. The humble place guaranteed to class enemies in socialist society was determined by their natural character—the need to avoid work and exploit others for personal gain. Even occasional association with *kulaks* was sufficient to brand an individual as unreliable and suspicious, even if the interaction was limited. "Very serious actions have been taken to consolidate the work discipline of

22. A Romanian joke comments on the continual postponement by the Communist Party of the communist paradise: What is the horizon of communism? It is a line connecting two points that is constantly receding.
23. ZML MDP.mg 57.f.2 fcs, 53 .ö.e., 1950.July.29.
24. ZML MDP.mg 57 f. 2 fcs. 49 ö.e., 1952, December 11.

Dózsa cooperative farm. . . . Subversive elements, or those who tried to bring the *kulaks'* voice into the group or fraternized with them, have been removed from the membership. These were the ones who last summer hindered the group's work, or at the height of the season went to work for the *kulak.*"[25] Society would be reconstituted, aided by scientific principles ensuring that the hierarchy in the workplace and society at large reflected a truly just distribution of effort, founded on moral integrity and respectability.[26] With the proper classification in place, standardized wage scales could be devised and productivity rewarded.

Finally, it is important to note that, although the social hierarchy of socialism was calibrated according to the individual's ability and willingness to work, not all forms of work were accorded the same value or prestige. Obviously, industrial workers had greater prestige than agricultural workers. Those engaged in productive labor—usually understood as the production of things—were endowed with higher value than those employed in the service sector or commerce, such as existed at the time. Being landless in a rural community, or even being a cooperative farm member, did not necessarily garner party favor if a person performed dishonorable work, as the following anecdote reveals:

> A. J. is a cooperative farm member who is completely landless and not a party member. At the meeting mentioned above he encouraged the members to leave the cooperative since they can't earn anything and their living is not secure. I determined in relation to the forenamed that he is a lumpenproletarian, who usually performs as a musician at parties and weddings in neighboring communities. He is used to earning his money easily and without honorable work, and this is why he tried to influence cooperative farm members.[27]

Indeed, A. J.'s willingness to speak out against the party's policies is a sure sign of his being an enemy of the people. I would also draw your attention to the thinly veiled reference to A. J.'s being a Roma (Gypsy) musician. Obviously, the socialist state categorized citizens according to Hungarian racial theories of diligence, not just class.

In ways eerily familiar from the 1920s and 1930s, party/state officials devoted themselves to extensive campaigns of "folk education" to reconfigure the social and

---

25. HBML XXIII. 9./b., 430 d., 340 sz., 1952, February 19.

26. The actual fate of class enemies during the Stalinist period is far more complex than official pronouncements suggest. As James Mark's (2005a, 2005b) research on the urban bourgeoisie during the 1950s demonstrates, so-called class enemies could and did find ways of participating in the new socialist economy and bureaucracy. In some instances, the party/state relied on their skills and expertise; in others, untrustworthy elements could reform themselves through engaging in physical labor. In village communities, who was defined as a class alien and who was punished for his or her social identity varied tremendously from place to place and party official to party official. For an example of the other extreme, that is, a former impoverished villager being branded as a *kulak*, see Lampland (1995, 144).

27. ZML MDP.mg 57 f 2 fcs 64. ö.e 1950, January 18.

behavioral landscape of laboring. This was a demanding chore. In Stalinist parlance, this was called agit-prop, and it was backed up by a more actively violent state apparatus than existed prior to 1948. National and county-level bureaucrats never ceased to underscore the centrality of these activities to local leaders, not least because local party representatives were not always the most willing participants in these endless tasks. "In order for true socialist work discipline to develop, we must first show the most class-conscious example. Work discipline will only improve if ideological education does not cease for one minute, if communists are properly divided into brigades and work groups. They should constantly train cooperative members at their work site and wage a battle against those subverting work discipline."[28] Communists were to set an example for the rest of the community by being good workers, as well as taking measures against those who resisted the new standards of diligence demanded by the party/state. More eager or audacious officials erred by attacking too many villagers, confusing the upright working peasant (*dolgozó paraszt*) with the evil *kulak*. These swings toward what in Marxist-Leninist parlance were called left-wing deviations in village politics had to be reined in because they tended to unite villagers against the party rather than permitting local party officials to exploit internal divisions within communities against one another. Transforming the political consciousness of reliable villagers entailed severing bonds of kinship, affection, and familiarity between local elites and their poorer peasant neighbors. It also required making clear to the working peasantry the cost of the socialist promise, that is, that one had to take sides against his or her neighbors to partake of the new rewards of socialist development. "If this offends them, then that is our mistake; we didn't transform their consciousness. A truly great life is waiting for the working peasantry, but they must do their part according to their ability."[29] Everyone would be sorted out in the end, and those who resisted the state's lead could find themselves excluded from the bounty of socialist wealth to come.

Engineering Instinct

What is the point of this exercise? Beyond recounting a curious tale about the moral connotations of social classifications, I have been intent on interrogating two central issues: (1) the tension between standardization and social engineering and (2) the physical embodiment of social qualities. I have argued that the standardization of workers did not constitute a process of deskilling to a simple common denominator. On the contrary, standardizing workers entailed complex techniques of differentiation, techniques carefully crafted in work science, occupational psychology, and scientific management. Indeed, it was the task of these fields to ascertain just how to differentiate between workers' aspirations and their abilities. Scientific pro-

28. GYMS MDP.mg 30 f. 2 fcs. 2 ö.e. 31 sz. 1952, January 27.
29. HBML MDP XXXII. 41 f. 2 fcs. 94 ö.e. 1949, September 6.

cesses of close observation, experimentation, and complex calculations were used to figure out who fit where in the standardized schema while, at the same, time social engineers tried to understand the degree of malleability available when adapting workers to new standardized protocols. Developing standardized templates for the labor force required social engineers to balance carefully their need to create workers suited to new productive tasks with their belief that scientifically calibrated standardizing protocols made explicit the already existing (and probably unchanging) features of workers. Social engineering was not a grand utopian adventure discovering new qualities in the labor force but a sustained process of scientifically vetting long-existing hierarchies of privilege and exclusion.

My analysis of class stratification and the standardization of laborers has also been informed by the wide-ranging critique of race and ethnicity in recent years, a critique that encourages us to look beyond ideologies of biologically determined features to examine the cultural, social, and moral elements intrinsic to racial and ethnic classification (e.g., Briggs 2003; Dominguez 1994; McCallum 2005). In this critique, racialization weaves the physical and the moral in a complex tapestry, whose warp and woof intertwine the fixed and fluid in equal measure.

> Racism is commonly understood as a visual ideology in which somatic features are thought to provide the crucial criteria of membership. But racism is not really a visual ideology at all; physiological attributes only signal the non-visual and more salient distinctions of exclusion on which racism rests. Racism is not to biology as nationalism is to culture. Cultural attributions in both provide the observable conduits, the indices of psychological propensities and moral susceptibilities seen to shape which individuals are suitable for inclusion in the national community. . . . If we are to trace the epidemiologies of racist and nationalist thinking, then it is the cultural logics that underwrite the relationship between fixed, visual representations and invisible protean essences to which we must attend. (Stoler 1992, 521–22)

Class and gender are equally implicated in these processes.[30] In this way, immutable bodies come to stand for immutable moral qualities, even though the process whereby bodies are invested with specific psychological propensities and moral susceptibilities is historically contingent and changes over time. Moreover, moral features are not just historically contingent; they also vary with an individual's social position.

Hungarians in the mid-twentieth century shared a cultural logic—or what, following Collins (1990), could be called a matrix of domination—justifying inequality. Class stratification was premised on fixed features of embodied skills and moral qualities, paired in curious (Lamarckian) fashion with a belief in the plasticity of social behavior. Work scientists sought to reveal inherent abilities and gauge their malleability through experiment and observation, insights that guided the

---

30. See the literature on the analytic imperatives of intersectionality, for example, Collins (1990).

procedures for occupational standardization. In fact, however, scientific processes committed to discovery were constitutive of the very social features and behavioral dynamics to be codified in standardized protocols. The moral valence of these practices, on the other hand, was not dictated by work scientists. Throughout the twentieth century, who deserved to be where in the social hierarchy was contested, no matter how firmly the powers that be sought to repress alternatives. Thus, when new political forces took power after 1947, interpretations explaining inequality changed overnight, even though the means of achieving fairness relied on techniques developed to justify an entirely different social structure. Only rarely, as in the story told here, do the ostensibly inherent qualities migrate so rapidly from one category of social bodies to another, shifting moral polarities while holding fast to the scientific mechanisms of classification.

A final note about contemporary uses of human nature and biology in economics and the social sciences more broadly. The quest for a rigorous understanding of human behavior has continued unabated since the 1930s, ranging broadly across the fields of psychology, management, economics, sociology, anthropology, cognitive science, genetics, and biology. We are currently enjoying a renewed fascination with genetic determinism, in both scientific circles and popular culture (Lancaster 2003). The trope "it's genetic" is heard in casual conversation as well as in debates over social policy. How are we to make sense of the deployment of claims about human nature and natural forms in the current climate? This can be explained not only by the strides in genetics in recent decades but also in terms of a broader social history about conservative politics in the United States. The disproportionate distribution of wealth is explained as the normal consequence of the international division of labor. Asian women's delicate hands assemble computer parts more effectively than the North American male; Third World migrants are "inclined" to carry out tasks shunned by European workers, whose skills lie elsewhere. These claims about human nature and historical necessity are quickly mobilized when criticisms about the social consequences of neoliberalism are raised. Markets are neutral and fair, neoliberal advocates explain, because the science of economics carries none of the political biases or social prejudice we can expect of governmental regulation. My analysis of work science and allied fields of business economics should erase this false dichotomy between the unreliable politics of states and the pure integrity of science. Investigating work science as a crucial feature of modern political economy and studying political economy as constitutive of scientific practices refute these claims. After all, if it is so easy to see the convoluted twists and turns that theories of instinct and human nature traveled in the not-so-distant past, and with such frightening consequences, it behooves us to bring all the strengths of our critical apparatus to interrogate claims made in the present that bolster inequalities based on theories of human nature—in science, in policy, and in everyday life.

■ EPA to Kill New Arsenic Standards

Whitman Cites Debate on Drinking Water Risk

ERIC PIANIN AND CINDY SKRZYCKI, WASHINGTON POST
STAFF WRITERS

Washington Post   March 21, 2001
(Edition F) Section A

The Environmental Protection Agency announced yesterday it will revoke a Clinton administration rule that would have reduced the acceptable level of arsenic in drinking water, arguing the evidence was not conclusive enough to justify the high cost to states, municipalities and industry of complying with the proposal.

The measure was one of dozens of environmental regulations and rules issued in the final months of Bill Clinton's presidency that were put on hold after President Bush took office Jan. 20 and subjected to further review.

In ordering the withdrawal of the standard, which was to have taken effect Friday, EPA Administrator Christine Todd Whitman said there is "no consensus on a particular safe level" of arsenic in drinking water. She rejected the arguments of environmentalists that studies show the limits are essential to protecting millions of Americans from cancer and other health threats, and suggested that a lower standard might be more appropriate.

"I am committed to safe and affordable drinking water for all Americans," Whitman said. "When the federal government imposes costs on communities—especially small communities—we should be sure the facts support imposing the federal standard."

Meanwhile, the Interior Department's Bureau of Land Management will announce today a proposal to suspend new hard-rock mining regulations for public lands that would have toughened environmental standards and made it more difficult for mining companies to escape financial liability for violations of anti-pollution laws, an Interior Department official said. The new rules—adopted Jan. 20—have prompted four lawsuits from mining groups, and bureau officials say it makes more sense to shelve the regulation while the disputes are settled.

These and other decisions in recent weeks provide a strong signal that Bush has embarked on a radically different course on the environment, energy production and federal land use than was taken by Clinton—a course that was not fully outlined during last year's presidential campaign

A week ago, Bush announced that he would not seek reductions in carbon dioxide emissions at power plants, breaking a campaign pledge. Then administration officials signaled that they might consider a court settlement that would significantly scale back a Clinton rule placing a third of the national forests off-limits to logging and road construction.

"They are mounting a bigger assault on the environment and public health than any other administration or the Gingrich Congress did," said Philip Clapp, president of the National Environmental Trust, referring to former representative Newt Gingrich (R-Ga.), House speaker from 1995 to 1998.

At least 34 million people in the United States are exposed to drinking water supplies with elevated levels of arsenic that occur naturally or as a result of industrial pollution and mining operations, according to studies by environmental groups. A National Academy of Sciences study concluded that arsenic in drinking water can cause cancer of the lungs, bladder and skin.

Experts say the biggest problems are found in the Southwest, Midwest and Northeast, while the Washington area has a relatively minor problem. The proposed rule, issued Jan. 22 after years of study, would have reduced the acceptable level of arsenic in water for the first time since the 1940s— from 50 parts per billion to 10 parts per billion.

Congress mandated a toughening of the standard, and the measure promulgated by the Clinton administration would have brought the United States into compliance with a standard adopted by the World Health Organization and the European Union within five years. Opponents and proponents differ sharply over the cost of implementing the rule, but state and local communities and industries would have had to spend billions of dollars collectively to replace filtration systems, switch to alternative water supplies and launch programs to educate the public on the dangers of arsenic.

Senate Budget Committee Chairman Pete V. Domenici (R-N.M.) praised the administration's decision to shelve the arsenic regulation, saying that it would spare states onerous costs.

"Communities faced with the daunting task of finding the money to adhere to the stricter standards can breath a sigh of relief," Domenici said. "I believe the rules rushed to print in January are excessive and based on insufficient science."

However, Erik Olson, a senior attorney with the Natural Resources Defense Council, which sued the EPA in February 2000 to issue the revised standard, said the administration's decision "will force millions of Americans to continue to drink arsenic-laced water."

"This outrageous act is just another example of how the polluters have taken over the government," Olson said.

To undo the regulation, the EPA must launch an administrative procedure that includes a public comment period. "This is in line with what Congress intended," said John Grasser, spokesman for the National Mining Association. "Congress gave the agency more time to examine the science and costs and the last administration chose, instead, to rush it out."

■ Arsenic and Water Don't Mix

San Francisco Chronicle
Editorial   August 27, 2001 (Final Edition)

California should set its own safer standard for arsenic in drinking water.
Let's not rely on President Bush to agree to an updated level of protection at
the national level.

The Legislature is halfway toward ordering a more stringent limit on
arsenic exposure, to be effective by Jan. 1, 2003. The state Senate has
approved the step (SB463 by Sen. Don Perata, D-Oakland). The measure is
to go before the Assembly Appropriations Committee next week.

The state's need to establish its own rule on how much arsenic we drink
was demonstrated by the Bush administration's blocking a toughened rule
put in place by President Clinton as he prepared to vacate the White House
last January.

Revision of the lax federal standard was shamefully overdue. The 50 parts
per billion arsenic level it allows dates back to 1942—before it was known
that arsenic causes several kinds of cancer and, at 50 ppb, poses a high
cancer risk of one in 100.

Both houses of Congress have voted to cut the level of allowable
arsenic—to the Clinton-ordered 10 parts per billion in the House measure,
or to an unspecified Environmental Protection Agency standard in the
Senate bill (approved 97 to 1). The two versions must be reconciled, and
the effort could face a Bush veto, anyway.

So Californians need assurance of Bush-proof protection from high levels
of arsenic in some water supplies, which can be corrected by controls on
harmful runoff and by water treatment. Some arsenic in water comes
naturally from rock formations.

The state should have its own law. The Perata bill does not specify an
arsenic standard, but sets up a mechanism for arriving at one and
publicizing it, through the Office of Environmental Health Hazard
Assessment. Sacramento's prompt enactment of the measure would send a
pointed message to Washington.

■ EPA Wants Review of Sulfur Standards

Whitman Postpones Clinton Rule Requiring Clean-Burning Diesel Fuel

Houston Chronicle   April 05, 2001 (3 Star Edition)
Section: Business
Source: Bloomberg News

WASHINGTON—The Environmental Protection Agency says it wants an independent study of a new rule that requires refiners to remove 97 percent of pollution-causing sulfur in diesel fuel, a decision that pleases refiners and farm groups.

EPA Administrator Christie Whitman said last month she'd uphold the rule "to protect public health and the environment," and has touted this commitment in defense of decisions to delay standards to reduce arsenic in drinking water and reject a 1997 global warming treaty.

The American Petroleum Institute, which represents the nation's oil and gas industry, asked Whitman for an independent review, arguing the rule was rushed through in the final days of Bill Clinton's presidency without regard to its impact on the diesel fuel supply.

The study would be "considered a major step backward" by automakers and environmental groups, said Paul G. Billings, a lobbyist with the American Lung Association.

"This is an effort to create a record to delay the rules," he said. For Whitman "to throw a bone to the oil industry is very disappointing."

The rule would affect refiners, automakers and drivers of heavy-duty trucks and buses that rely on diesel fuel. It aims to cut 2.6 million tons of smog-causing nitrogen oxide emissions each year, preventing thousands of asthma, bronchitis and other pollution-induced illnesses, and an estimated 8,300 premature deaths per year, according to the EPA.

In issuing the rule, Carol Browner, then EPA administrator, said its impact would be equivalent to eliminating air pollution created by 13 million of the 14 million diesel trucks and buses on the road.

An EPA spokeswoman said the agency plans to enforce the new standard and wants a third-party group, such as the National Academy of Sciences, to review the refiners' progress in making the equipment changes necessary to meet the standard when it goes into effect in 2006.

Clinton signed off on a series of environmental protections during his final days in office. President Bush's administration has stayed some for review and overturned some. Prohibitions on new hard-rock mining sites and development in 58 million acres of roadless wilderness have been abandoned, along with the arsenic standards and the U.S. commitment to the Kyoto global warming treaty.

In most cases, the administration has said the requirements were too costly for consumers and industry and risked damaging the U.S. economy.

"The Bush administration is running an environmental policy based on economics," said Ethan Siegal, who tracks political events for institutional investors as president of Washington Exchange. "I don't think anybody should be surprised."

Oil groups such as API opposed the diesel rule during two years of hearings and public comment, and API is now is suing to delay it. API's president, Red Cavaney, wrote Whitman on March 1 to say the group is pleased with her decision to have a study.

API says the costs of new engines and refining technology would drive small refiners out of business, causing fuel shortages and price increases.

The oil and gas industry was a major contributor in the 2000 election campaign. The industry gave $32.7 million, a 32 percent increase over the $24.8 million it gave in 1996, according to the Center for Responsive Politics, which tracks campaign finance. Republicans received about 79 percent, and former oil man Bush was the largest recipient, at $1.85 million.

METADATA STANDARDS

Trajectories and Enactment
in the Life of an Ontology

Florence Millerand and Geoffrey C. Bowker

■ There has been an explosion of scientific data over the past twenty years as more and more sciences deploy remote sensing technologies and data-intensive techniques such as magnetic resonance imaging (MRI). Just as we are increasingly monitoring and surveilling one another, we are increasingly tracking the processes of environmental change on every scale from the cell to the ecosystem. With the development of cyberinfrastructure (or, in its kinder, European coinage, e-science), we are exploring the possibilities of working with data collected in multiple heterogeneous settings and using sophisticated computational techniques to ask questions across these settings. Again, the parallel with our forms of social control is uncanny (see Foucault 1991; Luke 1999). The big questions of the day cannot be answered, it is claimed, without this facility; data from one discipline cannot build a picture of species loss that can inform policy, just as data from one source cannot profile a population in order to discipline it.

We face the question, therefore, of the preservation of, access to, and sharing of scientific data (Arzberger et al. 2004). In the traditional model of scientific research, data are wrapped into a paper that produces a generalizable truth—after which the scaffolding can be kicked away and the timeless truth can stand on its own. There has been relatively little active holding of very long-term data sets and little data reuse. In the current context, to the contrary, there is an emphasis on

The authors would like to thank our collaborators at University of California, San Diego, Karen Baker and David Ribes, both members of the Comparative Interoperability Project, SES0433369 Interoperability Strategies for Scientific Cyberinfrastructure: A Comparative Study, 2004–2007. For further information and publications visit: http://interoperability.ucsd.edu (there is a special prize for the millionth visitor). We also thank the LTER community participants for their interest in and openness to this research.

what is oxymoronically called raw data—which can be gathered together, analyzed, visualized, and theorized about to produce new syntheses. This is particularly so for the case of environmental data (Franklin, Bledsoe, and Callahan 1990), for which theories need to range over multiple temporal and spatial scales. It has been traditional in ecology for individuals or small groups to collect data in short-term projects (the length of a funding cycle) over small areas (1 square meter) (Lewontin 2000); this is no longer sufficient to the task (Bowker 2006). Ecosystems change in larger chunks of time (indeed, they follow multiple complex rhythms) and over wider areas of space than was traditionally conceived (O'Neill 2001). Thus, researchers need to be able to use data sets constructed by others and for different purposes; and they need to be able not only to reach some kind of ontological accord between the disciplines (allowing kinds and classifications to be shared) but also to be able to trust data produced by others—the traditional "invisible college" (Crane 1972) becomes a teeming city with multiple linguistic communities.

One of the major challenges for the development of a scientific cyberinfrastructure aiming to foster collaboration and data sharing through information networks is to ensure the frictionless circulation of data across diverse technical platforms, organizational environments, disciplines, and institutions. Or, to use the term of the art, to ensure interoperability. A central problem here is that the storage of, access to, and evaluation of the validity of data are extremely dependent on the ways in which the data have been collected, labeled, and stored. Although it may be possible for two colleagues in a discipline to share information about their data with a simple longhand note, there is unquestionably a need for more documentation in the case of pluridisciplinary teams working across multiple sites and scales. To deal with long-term questions, for example, a given data set may have been collected in one context using a homegrown set of protocols, often deploying outdated instruments and terminologies. The task of making those data available across disciplines and over time is, in general, an unfunded mandate—it requires a special kind of altruism to carefully code your data in ways beyond what you need for their immediate use. It is easy to see, therefore, why assorted technofixes are being discussed, debated, and to some extent deployed to address this problem set.

The resulting standards—most often conceived as being simple technical solutions—are being developed to permit the interconnection of systems and, thence, the free flow of data. The capacity for distributed, collective scientific work practice is posited on the existence of shared information infrastructures and collaborative platforms. These, in turn, require some base of shared standards. Although they are largely ignored and invisible (buried in an infrastructure or wrapped in a black box), these standards nonetheless constitute the necessary base for distributed cognitive work. In order to understand the modalities of collaboration in collective work—scientific and other—we need to understand standards.

In particular, we need to understand the forms and functions of metadata (data about data) standards (Michener et al. 1997).

Here we examine an infrastructure development project for an ecological research community—the U.S. network for long-term ecological research (LTER)—which is endeavoring to standardize its data management through the adoption of a shared metadata standard called the Ecological Metadata Language (EML). This is one of a suite of Extensible Markup Languages (XMLs)—there is Virtual Reality Markup Language (VRML) and even another EML, Educational Modeling Language (EML$_2$). This standardization process began in 1996, at the level of the LTER network. It crystallized in 2001 with the adoption of the EML standard by the community. It has since been the subject of a controversy that can be characterized as mission successful (by completing the implementation cycle) or success to come (by staying the course).

It is precisely these divergent visions that are the object of this chapter. We do not want to know the success conditions for the implementation of an information management standard in an organization as much as know from what *time* and according to which *point of view* the success or failure of the implementation is judged. We draw on an ethnographic study of the community to explore the alignment of diverse trajectories (Strauss 1993) in standards development. We explore this process as one of the enactment of a standard.

The Development of Information Infrastructures at the Intersection of Social Worlds

As for scientific activity in general, the development of information infrastructures for the sciences requires the cooperation of a heterogeneous set of actors—in this case, domain experts, information technology (IT) specialists, informatics researchers, and funding agencies. There is no linear narrative to be told; "The time of innovations depends on the geometry of the actors, not on the calendar" (Latour 1993, 80). In other words, we cannot track a single life cycle (development, deployment, and death) but must pay attention to the diverse temporalities of the actors. This perspective allows us to better grasp how the existence and even the reality of projects vary over time, in line with the engagement or disengagement of actors in the development of these projects or objects. Thus, although a technical object may exist in prototype form, it can be considered more (or less) real only to the extent that certain groups of actors rally (or do not) to its cause (Latour 1993).

What interests us here is the point of contact between the different trajectories of the human actors as well as the nonhuman actors in the process of standardization at work in the heart of the LTER research community. What happens in this process that leads one group of actors to formulate an alternative history to the success story? Which trajectories interact with one another, and how do they adjust accordingly? How are certain trajectories redefined?

An immediate problem is to know which trajectories to follow (Fujimura 1991; Star 1991; Timmermans 1998) because the choice of following any one, in particular, over another can lead to a different understanding of the social, technical, and organizational configuration of the study. Infrastructure studies are a useful source here because they shine analytical light on rarely studied phenomena—such as the invisible work carried out in the background by actors whose performance is considered so much the better to the extent that it is self-effacing and invisible (Star and Ruhleder 1996; Bowker and Star 1999; Star and Bowker 2002).

We conceptualize the EML standard as a support for the coordination of different social worlds: domain researchers, standards development teams, and information managers concerned with its implementation. EML was defined a priori as a solution to a set of technical problems—a solution from which will issue the one good tool that can be used by all. But this technical standard and its implementation process, in fact, speak directly to the organization of scientific work; it assumes specific configurations of actors, tools, and data. In order to explore this dimension, we deploy here the concept of enactment, developed by Karl Weick (1979). In this tradition, Jane Fountain invites us to distinguish between an objective technology, that is to say, a set of technical, material, and computing components such as the Internet, and an enacted technology, that is to say, the technology on the ground as it is perceived, conceived, and used in practice in a particular context. In this view, the way in which actors enact technical configurations such as standards depends directly on their imbrication in cognitive, social, cultural, and institutional structures. Organizational arrangements (characterized by routines, standards, norms, and politics) mediate the enactment of technologies, which in return contribute to the refashioning of these arrangements.

We propose, therefore, examining enactment in action—to trace two sets of histories of a single process of standardization, by restoring the artifacts, actors, and narratives to the context whence they emerged. This perspective permits a better understanding of the social and organizational dynamics at the heart of projects for the development of large-scale information infrastructures.

The Long-Term Ecological Research Community and the Ecological Metadata Language Standard

The LTER program constitutes a distributed, heterogeneous network of more than 2,300 research scientists and students. Formed in 1980, the network currently consists of twenty-six sites, or research stations (Hobbie et al. 2003). Each is arranged around a particular biome—for example, a hot desert region, a coastal estuary, a temperate pine forest, or an Arctic tundra—in the continental United States and Antarctica. A twenty-seventh site is charged with the administration and coordination of the group. The program mission is to further the understanding of environmental change through interdisciplinary collaboration and long-term research projects.

One of the chief challenges of LTER is to move beyond the plot of traditional ecoscience to analyze change at the scale of a continent and beyond the six-year funding cycle or thirty-year career cycle of the scientist to create baselines of data spanning multiple decades. Although the preservation of data over time, and their storage in conditions appropriate to their present and future use, has always been a priority within the different sites of the LTER network, there has been a new urgency with the development of a cyberinfrastructure projects aiming to encourage data sharing across the community.

Each of the twenty-six sites in the network takes responsibility for the management of research data produced locally; and each in general has its own information system (its own databases). An information manager is charged with the development and maintenance of local infrastructures. Across the network, then, data are stored autonomously by the different sites—a fact that renders the search for and access to data relatively complex and laborious (which, in turn, naturally militates against the network's realizing its mission). Accordingly, a project was put into place in 1996 to initiate a networked information infrastructure permitting the federation of the local databases and, thus, data exchange.

The project has encountered three major challenges: the heterogeneity of the data that circulates through the research community; their wide dispersal; and the multiple systems of coding and storing (Jones et al. 2001). Beyond the diversity of data attached to a given scientific project, there can also be an extreme disparity in their organization and formatting, depending on the collection protocols adopted. For example, data to measure the amount of chlorophyll present in a sample of seawater may be organized into separate files corresponding to the number of trips made, whereas the same measures taken over a year in a given lake may be held in one single file. Further, local cataloguing cultures generally use information (or metadata) that is not necessarily understandable outside a given research project, site, or discipline. Thus, special (personalized) measurement units can be created for the analytic needs of a particular research project (e.g., the ungeneralized unit "number of leaves per change of height in a plant"). In this context, the LTER network office soon saw the need for a set of data standards. Or, to be more precise, the need for standardized methods for the collection of metadata has been taken as the preferred solution for data interoperability.

In an ideal world, the metadata contain all the details necessary for all possible secondary users of a data set (an ideal solution that evokes Baruch Spinoza's problem—to know a single fact about the world, we need to know every fact about the world). These include detailed and diverse information, such as the names of the researchers who collected the data, the title of the project they were working on, the project summary, the key words, the type of biome, the sampling techniques, and the calibration of the measuring tools at the time of data collection. By extension, the possibility of complex analyses drawing on physical, chemical, and biological data within a given geographical area depends on the quality of the metadata. (Let us note at this point that metadata are not necessarily the only possible solution; however, they have been presented—organizationally and intellectually—as central to this community.)

Although they appear to be mere technicalities to some, metadata take on a central importance in the production of scientific theories in the degree to which they condition access to data, guarantee their integrity, and delimit their interpretative uses. When we are dealing with comparative or long-term studies, we find that material conditions change over time and space—instruments might become more accurate, for example, which indicates the need for the precise documentation of their calibration. When combining data across the disciplines, metadata do more than provide a convenient label; they structure the conversation that ensues. To use an analogy with normal language use, there are communities for whom *casualty* means "someone injured or killed" and others for whom it is simply an euphemism for "fatality." Or there are some for whom *democracy* means "rule of the people" and others for whom it is a euphemism for "capitalism." Unless we can calibrate across the communities (and it is often a difficult act of imagination to recognize how local and specific our use is) then we cannot communicate.

The EML, a metadata description language, is precisely a standardized language for the generation of metadata in the specific domain of the sciences of the environment—it is much better to use *sciences* (in the plural) here because each domain has its own configuration of classifications, instruments, dates, and places. EML is the standard that the LTER community has adopted when it is engaged in the process of standardizing its scientific data management practices.

Controversies

The process of standardization that the LTER community is engaged in has two major objectives: the promotion of interdisciplinary collaboration through data sharing and the improvement of long-term data preservation (Hobbie et al. 2003). Although both these objectives and the deployment of EML to their end are generally agreed on, conflicting voices could be heard at the moment of deployment.

Here we examine two narratives, from two categories of actors, that tell radically different tales about the EML standard as successfully implemented or as still a work in progress. The first comes from the developers of the standard, who include the experts who wrote the specifications for EML together with the coordinators of the LTER network. The second presents the point of view of those enacting the standard—that is to say, the information managers whose task it is to implement it at a given site. At the time of this study, the success narrative was carrying the day—it was already formalized, written up in reports, whereas the second one was diffuse and oral. (As, in the context of policy work, in John King's dictum, "some numbers beat no numbers every time," in the context of computer science funding, written beats oral every time.)

With respect to methods, interviews were conducted with both groups of actors, and participant observations were carried on throughout the standard implementation at two of the LTER sites. Detailed document analysis was systematically performed.

Narrative 1: "EML Is a Success: The Entire LTER Community Has Adopted It"

Version 1.0 of EML first saw the light of day in 1997 at the National Center for Ecological Analysis and Synthesis (NCEAS) in Santa Barbara, California. It was the product of a researcher in ecological informatics working with two doctoral students. EML responded in the first instance to internal preoccupations within NCEAS, which since its creation in 1995 has been addressing the absence of tools and techniques for analyzing and synthesizing environmental data. A grant was submitted to the National Science Foundation, which funded its development.

Technically, EML is based on two emergent standards (Standard Generalized Markup Language, SGML, and XML) that were, *grosso modo*, developed to turn networked information from simple text fields into searchable, combinable databases. Its content is drawn from the main data description types in use in the domain—such as those recognized by the Ecological Society of America, itself a pioneer in preserving data sets alongside papers. Versions 1.0 to 1.4 cascaded out between 1997 and 1999. They were tested within NCEAS. Given the difficulties encountered in use, a major revision of the language was suggested (would that we could vary natural languages so simply)—and a second grant proposal was written and subsequently funded. The development team went from being three people to a collaboratory (Olson et al. 2001)—a collaborative platform based on voluntary participation and open to the whole community of environmental scientists. This open development model was not immediately successful, even though the team was able to attract some more developers, including—for the first time—a separately but synergistically funded information manager from the LTER. The development of EML went on apace, with several significant structural changes being made. Seventeen versions were produced between 1999 and 2002.

In 2001, the team reckoned that it had produced a stable version of EML (version 1.9, which was accordingly anointed with the title EML 2.0 beta). The team presented its product at the annual conference of LTER information managers, held that year in Madison, Wisconsin. Discussion was lively, but responses were extremely favorable—the information managers recognized the usefulness of such a standard for the LTER community and were moved to formally adopt EML. Version 2.0 was put into circulation; the LTER network scientific community (one of the most important communities in environmental research) adopted the standard. In short, the EML project was a resounding success.

Narrative 2: "EML Is Not (Yet) a Success: It Needs to Be Developed Further Before It Can Be Used"

At the period of the creation of EML in 1997, the LTER network sites already had in place systems for managing their scientific data. Depending on the site, these systems were more or less formalized—that is to say, they did not necessarily use the same standardized vocabularies, even if some of them, broadly speaking, used the data descriptors recognized by the American Society of Ecology—leading to the standard problem of almost compatibility. In 1996, the inauguration of a project to develop a network-wide information system stimulated discussions about standardizing data management procedures and encouraged the development of a common tool set for the information managers of the community. But there was still no central initiative covering the whole network. In 2001, the EML project received a favorable reception from the information managers, who by consensus adopted it. The implementation began.

Although some sites began the work of implementing the standard relatively quickly, most of them ran into significant problems. The standard is complex, and it is difficult to understand in its entirety. The technical tools intended to facilitate the implementation of the standard proved unusable, that is, incompatible with existing local practices and infrastructures. And, in general, there was just a huge amount of work to be done (on top of the normal workload) with minimal resources—some sites had to undertake a complete restructuring of their data management practices.

Numerous ad hoc solutions were brought to bear—for example, homegrown tools that some of the information managers shared among themselves to facilitate the work of the conversion of local systems into the format required by the EML standard.

The information managers organized two workshops devoted to the implementation of EML in 2003 and 2004 in which the developers participated. These led to the production of a synthetic "best practices" document for EML implementation, which had a material impact on implementation at a number of sites. This, in turn, led to a five-step implementation plan formulated conjointly by the information managers and the LTER network coordinators.

At the annual conference of information managers in 2005 in Montreal, progress was seen as somewhat mixed—EML implementation was seen as a complex and laborious process whose outcomes in terms of improvement of data management remained somewhat difficult to identify. EML was not yet a success—it had to be partially refactored to be usable.

Trajectories

On the one hand, then, we have a success story about the EML standard, highlighting its adoption by consensus by the LTER research community. On the other hand, we have a very mixed picture—varying widely from site to site. Whereas the first story moved into the "happily ever after" phase in 2001, the second had barely gotten beyond "once upon a time."

A simplistic reading of these two narratives would say that the measure of success of a standardization process differs as a function of the different phases one is looking at (here, the phases of conception and development of the standard in contrast to that of deployment and implementation). In other words, we could say that the information managers cannot recognize the success of the project yet at this time—they will only be able to see it once the project is finally completed. This evolutionist reading of technology development projects unfolding in an objective time frame does not advance our understanding of what really happens during the periods of emergence, development, maturation, implementation, and so forth. Further, it continues to privilege the second story over the first—considering the success of the standard as being always already assured, with full confirmation coming in the natural course of events. Thus, it favors the invention of the standard over its innovation (Schumpeter 1934), that is, its deployment and enactment (Baker and Millerand 2007).

A temporal analysis of technology development projects should seek, rather, to account for their evolution in terms of the multiple temporalities into which they are integrated. It would then become possible to account, from the point of view of the actors, for the whole set of events—including the more troubled periods when folks do not want to talk about it (e.g., it does not sound good in the next funding application) while others still seek to find a voice (e.g., because they are too low status to be heard, or if they are heard, they are not using a technical language that the developers understand).

Multiple Trajectories

It is striking the degree to which all the actors involved in the standardization process (EML developers, LTER network coordinators, information managers, domain researchers, and so on) have supported—and continue to support—the EML project. As already mentioned, neither metadata nor the EML standard was the only possible solution to ensure data interoperability through the network. Nevertheless, they all believe in the idea of a metadata standard permitting the exchange and sharing of data throughout the LTER network and beyond. In this sense, this is not a case of the imposition of a standard by one group of actors (developers and coordinators) on a hostile, resistant group (information managers or domain researchers). The latter have always been highly supportive of the project, up to including the status of EML implementation in the information management review criteria for the sites (Karasti and Baker 2004; Baker and Karasti 2005). It is at the moment of the actual implementation of the standard at a give site when critical problems emerge and discordant voices can be heard.

The recognition of these difficulties and the controversy that has ensued have contributed to bringing the status of EML as a usable standard into doubt. Two years after its adoption, an inquiry revealed that it had not yet been completely implemented in a single one of the twenty-six network sites and that the tools developed explicitly for this purposes remained largely unused. EML seemed to be a standard in name only.

The juxtaposition of these two narratives reveals the confrontation of two visions of the EML standard. An imaginary dialog, inspired by Bruno Latour's (1993) history of the Aramis project reveals the gulf between the two sets of actors:

> DEVELOPERS: "EML 2.0 exists—the bulk of the work has been done, all we need to do is implement it."
>
> INFORMATION MANAGERS: "All we need to do . . . !? But a metadata standard is just a language. No matter how perfect it is, it exists only if it's being used—if it serves, above all, to integrate data."

In other words, in 2001 the EML standard was a metadata standard without data.

We propose to read these differing perspectives on EML by restoring them to the trajectories of the actors concerned. Thus, from the point of its developers, the EML standard was, above all, one of research and development. The project goal was the creation of a standardized description language for metadata, not its materialization. Its ambition was to make itself the reference standard in environmental sciences. From this perspective, the development of the standard and its adoption by the wider research community of environmental scientists constitute the main success criteria for the project. From the point of view of the information managers, the EML standard represented a set of tools and practices for the better management of scientific data—notably by improving the quality of metadata produced within each of the sites. According to this view, the successful incorporation of this new tool—and the new modes of practice that accompany it—within local sociotechnical infrastructures constitutes the major success criterion for the EML project. Finally, from the point of view of the scientists belonging to the LTER research community, the EML standard is a technical tool that opens the door to multisite research endeavors through a better form of access to and sharing of data and that promises a better diffusion of data beyond the LTER network. From this perspective, the capacity to carry out multidisciplinary projects in very large data sets through a single interface constitutes the main criterion for the success of the project.

Trajectory Alignment

We read the implementation work carried out by the information managers as a process of appropriation (Millerand 1999) of the standard, in the course of which the work of trajectory alignment is done. The appropriation of the EML standard in the different sites worked out, in effect, as the adjustment of the technical tool to local contexts, and the adaptation of preexisting practices to new ways of working. Concretely, this entailed a real work of bricolage from the information managers seeking to incorporate a (generic) standard into a (local) context, which both gave it its purpose and permitted its use.

The trajectory of EML according to the first narrative was born of the main descriptors of ecological data in use in the domain, became a research and development project at NCEAS, and then was adopted as the metadata standard of reference for environmental sciences. It seemed to take a turn or a certain reorientation as it began to circulate in the LTER network. From that moment, the description of the EML project as one of conception, development, deployment, and implementation ceased to work. In the implementation phase, there was redevelopment work, which led to a reconsideration of the conceptual basis of the work, and then some more redevelopment for reimplementation, and so forth. The EML project was changing—the set of trajectories had to be realigned.

What happened then in this implementation phase of the standard that necessitated more and more implementation work? The information managers spontaneously responded that there was a lack of tools permitting the conversion of local metadata systems into the EML format. Equally, they complained that there was a weak understanding of real implementation processes from the network coordinators, who seemed to them to have unrealistic expectations. The problem was that data management practices are not solely dependent on the types of technical infrastructure; they are also, and above all, intimately linked to the nature of the research projects being studied and to the disciplinary and organizational cultures of the sites—in short, to the local structure of scientific work.

The following two interview extracts illustrate, on the one hand, the local and contingent nature of the scientific work being done and, on the other, the complexity of the information managers' task of cataloguing research data.

I was getting nutrient data, and my units came in as micromoles with the micron symbol and capital M, micromoles. When I started having to go into EML, which does not have that unit, I had to figure out, well, what actually is this unit. And in digging deeper and going to our lab that processed these data I found out it's not micromoles, it's micromoles/liter. And I am not a chemist so it just didn't mean anything to me. You know, I am just organizing and posting this type of data, and so it really opened my eyes that I have a bigger issue here than I thought, you know, because here we've got people reporting things as micromoles, which is not proper. But that is just the way the work is done, and shared, and no one ever questioned it. That's kind of interesting. So I started data set by data set trying to retrofit everything back into, you know, EML. And I have this ongoing list of these custom units that I am compiling, making my best guess at and then I am going either to the actual, you know, my PI or a collaborator that gave me those data, and having to sit down with them to say can you please verify, if you were going to describe this unit in EML as a custom unit does this make sense. Are you reporting it the proper way. Are you calling this the attribute what it would universally be called, that kind of thing. . . . (IM_L, interview, 2005, San Diego, Calif.)

This first interview extract provides an example of the locally situated work that does not yet scale as a joint measurement unit within the framework of the shared conventions of a community of practice—in this case, the LTER network. The retrofitting referred to is the occasion for a lot of cyberinfrastructure disasters—data that were understood well enough by a local group often have to be completely revisited to be understandable to a wider community. Designers persist in seeing the conversion task as being easy, but this assumes that the data being fed in are clean and consistent.

Micromoles Per Liter and Micromolar are measurement units for concentration. Technically, both are micromoles per liter, and so equivalent in magnitude. [But] their scopes are different, because micromoles per liter can be used for a particulate or dissolved constituent, and micromolar is correctly used only for dissolved. So they are not exactly interchangeable.

Micromoles Per Liter and Millimoles Per Cubic Meter are equivalent in magnitude, but different disciplines have preferences for one or the other.

[Also], if you happen to be in open ocean, you would run into micromoles per kilogram and micromoles per cubic meter, which are similarly equivalent only at sea level, because interconversion depends on pressure. . . . (IM_M, interview, 2005, San Diego, Calif.)

This second interview extract gives an example of measurement units that are a priori identical but that mean different things in different disciplinary environments.

Taking a step back, the general problem can be characterized thusly. In the grand old days of the nineteenth through early twentieth centuries, when scientific certainty was at its zenith, it seemed as if there was a clear and consistent classification of and hierarchy among the sciences. The most famous example is Auguste Comte's classification of science into a classificatory tree going from mathematics through physics and chemistry down (or up, depending on your inclination) to sociology. Each part of each discipline was divided into statics and dynamics. This was also the period of the discovery of the principle (not, be it noted, the fact) of division of labor; Charles Babbage mirrored his computer on factory production techniques—making a complex task easy by splitting it into a set of serial subtasks. Together, the classificatory principle and the division of labor created a picture of scientists as workers in a giant collective enterprise—in Jules-Henri Poincaré's terms, workers lay bricks in the cathedral of science. Some, indeed, saw the end of the period of "heroic science" as the principle of the division of labor emerged.

Nowadays we are running into the question of whether cathedrals need blueprints (Turnbull 1993). What could possibly guarantee that all these bricks will fit together into a seamless whole? There are two options—in close parallel with "blind watchmaker" positions. Either there is a higher entity (the universal scientific method and the positivist classification of science, in this case) ensuring that they all fit or there is a constant, contingent, local process of partial fitting and constant disordering that can still, in the long term, guarantee that at any one "join" there is a fit (although there cannot possibly be across the set of joins). Both scientists and information systems designers have been working largely from the former assumption. When they face the reality of the latter, they are constantly surprised—they have not been following the scientific method as faithfully as they thought (their databases are dirty; their units are ambiguous), and it does not all fit into clearly designated, separable chunks (two disciplines might both claim control over the same measurement unit).

The Case of the Dictionary as Articulation Strategy

At the start of 2005, with the tools created by the developers not yet being used by information managers and the implementation dragging on, the network coordinators began the development of a new tool intended to accelerate the implementation of the standard among the sites. In parallel—and partly in reaction to this project—some information managers initiated the development of a, in-house tool: a repertoire of measurement units.

One of the principal difficulties that the information managers faced was tied to the complexity of the work of translating their metadata into the EML language—notably with respect to measurement units. On the one hand, the dictionary of measurement units that came with the EML standard essentially cataloged physical measurement units of ecological phenomena—whereas most LTER network sites were using biological and ecological measurement units. On the other hand, it is extremely difficult to describe in a standardized language special or personalized measurement units—that is to say, units developed for some specific purpose in a research project and that only really make sense in the context of that project.

Faced with these difficulties, some information managers then began to exchange lists of measurement units (including local ones) used at their site, so as to compare their respective translations and to catch any inconsistencies. This quickly evolved into a project to transform these lists into an LTER-wide catalog of units. The plan was to produce a dynamic online tool available through the LTER intranet. The team, which until then had been made up solely of information managers, expanded to include a representative from the network office. They developed a prototype integrating the unit lists of six sites. This was presented in August 2005 to the annual conference of LTER information managers. It was represented both as an implementation aid for the EML standard and as an example of a successful collaboration between information managers and developers/coordinators.

Technically, this tool provided the information managers in the network access to definitions of measurement units in EML (including some specialized units) and allowed them to propose corrections to the EML unit dictionary and to add definitions of other units. But it did considerably more than facilitate the conversion from one format to the next—it was, above all, a work of social coordination.

These local ad hoc initiatives and the network-wide projects can be seen as strategies for re-articulating the work—on the one hand, among the information managers themselves and, on the other, between them and the developers/coordinators. The information managers did not need the tool itself as much as they needed all this work of information mediation that the enactment of the standard brought to bear: How best can we describe such and such a measurement unit? Is a given measure a local or a network one? Can this unit be added to the EML dictionary? The work of producing the dictionary of units became a tool for facilitating coordination and cooperation between different worlds. In other words, by creating a tool that could be used locally at each site and yet contributed to the improvement of the standard, the information managers created a boundary object (Star and Griesemer 1989) that was capable of supporting this articulation work between enactors and developers.

Enactment

We propose using the concept enactment to better understand the complexity of the work of the information managers. The information managers not only ensured the implantation of the EML standard and the programming of some additional functionalities that could make it operational (i.e., its implementation)—they also worked on its interpretation; that is to say, they worked on its staging (in the theatrical sense of the term), which involved coadapting the standard and local work practices. These mutual adaptations are better seen not as local resistances but as necessary adjustments without which the EML standard could not operate within the LTER community.

In what ways did the EML standard contribute to changes in the social worlds of the actors? And how did these worlds work to change the standards? These changes involved, first, the identities of the actors and their organizational structures and, second, the script of the technical tool.

New Organizational Roles and Structures

Throughout infrastructuring in general and this standardization process in particular, the information managers as a community of practice became visible within the LTER network (Karasti, Baker, and Halkola 2006). In the same way, certain aspects of their work that, until then, had been little known and recognized became explicit. In 2004, when there was an efflorescence of in-house tools in particular sites and the information managers were being integrated with the developers, they achieved "official" status as developers—that is to say, they appear on the list of credits attached to the EML documentation. The complexity of their task (taking the local and rendering it into a universal language—a task that even the engineers of Babel found daunting) was recognized.

That said, even though the transformation of organizational models within the LTER community forced a reorganization of working patterns (from site-based to a federated structure) and even though the role of information managers was considerably transformed, their status as technicians whose task is to provide support and maintenance remains dominant in the network—notably among the domain scientists. Furthermore, even though the team of experts has recognized their contribution as developers of the standard, their contribution remains ambiguous to the extent that developers retain the tendency to see initiatives from the information managers as being too local and not state of the art. Thus, the synthetic "best practices" document produced by the information managers was judged to have too many signs of its origin within the LTER community to be integrated into the documentation of the standard. (And, we hesitate to say "of course," of course, but as a matter of course the social and organizational innovation was similarly not included in the documentation of the standard).

It remains the case that the set of actions carried out by the information managers during this standardization process revealed that another organizational configuration was possible—if only at the very basic level of resource allocation. If the development of a metadata standard for a research community such as the LTER requires significant funding, then so a forteriori does its enactment within a given setting. More concretely, the information managers contributed to putting into place a new representative structure—in this case, a permanent committee formed equally of information managers and domain experts, whose mission is to ensure a representative and advisory role in the development of integrated network information management practices—the Network Information Advisory Committee (NISAC). This committee came into being one year after the adoption of the EML standard (a somewhat lengthy gestation period!), when the initial sets of difficulties forced the information managers to initiate a dialog with the domain scientists. Out of this committee came the plan to implement the standard in different stages.

Beyond this new organizational structure, a new form of collaborative work at the intersection of local, site-based work and global (network) activity was experimented with successfully. The project of building a dictionary of measurement units constituted a veritable innovation (from below) within the LTER community, to the extent that, on the one hand, it opened the way for the transformation of a local initiative into a project for participatory design at the network level and, on the other, created a new collaboration space between two groups of actors that had not been directly associated before (information managers and developers/co-ordinators).

Redefinition of the Standard

The standard itself has been the object of multiple versions over the course of its development (in part as a result of the new members integrated into the team), but what was presented in 2001 to the LTER information managers was a black box—a final version. The box was reopened, and not always in the same way, across the set of sites.

Thus, as we have seen, certain lacunae in the standard were identified and the implementation plan itself changed to accommodate the different rhythms of integration of the different sites. More generally, there are the kinds of difficulties that come into play when we try to enact a standard generic enough that it can in principle apply to any kind of data (Berg 1997). The EML project included a definition of a certain role for researchers—that of describing their data in this new language (Jones et al. 2001). Indeed, researchers have always been envisaged by the developers as the future users of the standard. They can both describe their data collections in EML terms and carry out complex, integrative studies using vast data sets as a result of these standardized descriptions. It is interesting to note that the role of the information managers has never been mentioned (at least not explicitly) in scenarios of EML use.

And yet, in practice, several LTER researchers refused their role—principally through lack of time (standards tend to be an unfunded mandate) and interest. It should be recognized that the implementation of the standard can double the amount of work they need to do to enter their data. Moreover, the investment in time to learn the EML language (without mentioning that of learning the conversion tools) has constituted a point of no return for many. The information managers have taken on this role that was a priori destined for others.

We can certainly read this redistribution of roles as a coming from a transitional period—and thus imagine that the LTER network researchers will get up to speed with EML to the extent that the tools become easier to use and the standard is taken up within environmental science generally. But the question of training researchers to use EML (and, more widely, to understand the new forms of information management associated with an integrated infrastructure) is at present hanging—and it seems likely that the information managers will continue to pick up the slack.

Conclusion: Infrastructural Changes

In one sense, then, the EML standard did not change anything. The division of labor remains the same (information managers are still in charge of the production of EML metadata), roles are stable (information managers contribute to the redevelopment of a standard for the LTER network while developers work on the development of a standard for environmental science in general), and local practices are confirmed (information managers share ad hoc solutions and in-house tools). Yet the EML standard has changed the world. The actors' identities have changed (information managers are recognized as developers and not merely implementers), new organizational structures have been built (information managers are now represented in the NISAC committee), and new forms of work are proposed (a collaborative space between the sites and the network has opened up).

The implementation of EML is not simply a case of upgrading an existing system. It consists, above all, in redefining the sociotechnical infrastructure that supports this articulation of technical, social, and scientific practices. These redefinitions have significant consequences socially and organizationally. Because the tools are intimately imbricated in local work practices and because the EML standard operates only within a given (social, technical, and organizational) configuration, its enactment requires infrastructural changes.

This is why it does not make sense to see standards simply as things out there in the world that have a predetermined set of attributes. In information systems, standards are in constant flux—they have to migrate between communities and across platforms. Closure is a narrative that serves a purpose, not a fact that describes an event.

We slice the ontological pie the wrong way if we see software over here and organizational arrangements over there. Each standard in practice is made up of sets of technical specifications and organizational arrangements. As Latour (2005) reminds us in another context, the question is how to distribute qualities between the two—what needs to be specified technically and what can be solved organizationally are open questions to which there is no one right answer. By assuming that specifications can exist outside organizational contexts, we have already given the game away, leading to the continual need for the kind of innovation detailed in this chapter. And the innovation is always forgotten because the same ontological mistake—made elsewhere, by other people—next time will again occasion its necessity. Indeed, a test for ontological errors in general is that, although we can say the same thing a hundred different ways over a span of years, there is no way the message can be heard *until* the organizational changes have taken place such that reception is possible (Douglas 1986). Both standards and ontologies (the one apparently technical and in the realm of machines, the other apparently philosophical and in the realm of ideas) need to be socially and organizationally bundled— not as a perpetual afterthought but as an integral necessity.

■ Horse's Ass

The following is an apocryphal story, but none the less, amusing story on the permanency of bureaucracies.

The U.S. standard railroad gauge (distance between the rails) is 4 feet 8.5 inches. That's an exceedingly odd number.

Why was that gauge used?

Because that's the way they built them in England, and English expatriates built the US railroads.

Why did the English build them like that?

Because the first rail lines were built by the same people who built the pre-railroad tramways, and that's the gauge they used.

Why did 'they' use that gauge then?

Because the people who built the tramways used the same jigs and tools that they used for building wagons, which used that wheel spacing.

Okay! Why did the wagons have that particular odd wheel spacing?

Well, if they tried to use any other spacing, the wagon wheels would break on some of the old, long-distance roads in England, because that's the spacing of the wheel ruts.

So who built those old rutted roads?

The first long-distance roads in Europe (and England) were built by Imperial Rome for their legions. The roads have been used ever since.

And the ruts?

Roman war chariots first made the initial ruts, which everyone else had to match for fear of destroying their wagon wheels and wagons. Since the chariots were made for, or by Imperial Rome, they were all alike in the matter of wheel spacing. The Imperial Roman war chariots were made just wide enough to accommodate the back ends of two war horses.

Thus, we have the answer to the original question. The United States standard railroad gauge of 4 feet, 8.5 inches derives from the original specification for an Imperial Roman war chariot.

And now, the twist to the story . . .

Source: http://www.pluggednickel.com/home/picayunepursuits.html (December 11, 2007).

When we see a space shuttle sitting on its launch pad, there are two big booster rockets attached to the sides of the main fuel tank. These are solid rocket boosters, or SRBs.

Thiokol makes the SRBs at their factory at Utah. The engineers who designed the SRBs might have preferred to make them a bit fatter, but the SRBs had to be shipped by train from the factory to the launch site. The railroad line from the factory had to run through a tunnel in the mountains. The SRBs had to fit through that tunnel.

The tunnel is slightly wider than the railroad track, and the railroad track is about as wide as two horses' behinds.

So, the major design feature of what is arguably the world's most advanced transportation system was determined by the width of a horse's ass! So, the next time you are handed a specification and wonder which horse's ass came up with it, you may be exactly right.

Specifications and bureaucracies live forever.

■ To Avoid Fuel Limits, Subaru Is Turning a Sedan into a Truck

DANNY HAKIM

The New York Times   January 13, 2004 (Late Edition—Final)
Section A, Business/Financial Desk
Dateline: Detroit, January 12

The Subaru Outback sedan looks like any other midsize car, with a trunk and comfortable seating for four adults.

But Subaru is tweaking some parts of the Outback sedan and wagon this year to meet the specifications of a light truck, the same regulatory category used by pickups and sport utilities. Why? Largely to avoid tougher fuel economy and air pollution standards for cars.

It is the first time an automaker plans to make changes in a sedan—like raising its ground clearance by about an inch and a half—so it can qualify as a light truck. But it is hardly the first time an automaker has taken advantage of the nation's complex fuel regulations, which divide each manufacturer's annual vehicle fleet into two categories. Light trucks will have to average only 21.2 miles a gallon in the 2005 model year. By contrast, each automaker's full fleet of passenger cars must average 27.5 miles a gallon.

The move will let Subaru sell more vehicles with turbochargers, which pep up performance but hurt mileage and increase pollution. "It was difficult to achieve emissions performance with the turbos," said Fred D. Adcock, executive vice president of Subaru of America. They also made it hard to meet fleetwide fuel economy standards for cars.

Subaru's strategy highlights what environmentalists, consumer groups and some politicians say is a loophole in the fuel economy regulations that has undermined the government's ability to actually cut gas consumption. The average fuel economy for new vehicles is lower now than it was two decades ago, despite advances in fuel-saving technology.

"This is a new low for the auto industry, and it would make George Orwell proud," said Daniel Becker, a global warming expert at the Sierra Club.

It is particularly striking that Subaru wants to call the Outback a light truck because many of its owners see the wagon version as a rugged alternative to a sport utility, and the Outback sells best in those parts of the country, like college towns, where many people think it unfashionable to own an S.U.V.

"I probably can't count my friends with Outbacks on one hand—I'd have to use feet and toes," said Elizabeth Ike, 29, a fund-raiser at Sweet Briar College, which is an hour south of Charlottesville, Va. She said "the Outback might as well be Charlottesville's official car," adding that the town "likes to think of itself as an island that is more globally aware than the rest of the state."

"I don't want to speak for my friends, but I think they probably don't want to be that person in the Excursion," she said, referring to Ford's largest sport utility.

Subaru, a unit of Fuji Heavy Industries, says the new Outback, which will make its debut next month at the Chicago auto show and go on sale this spring, will retain its not-an-S.U.V. image because the changes being made are technical in nature. What customers will notice will be the new Outback's glossier look, executives said. Further, the base model will be more fuel efficient than the current version.

They said that calling the Outback a light truck will also let them offer the option of a tinted rear window not allowed on passenger cars.

Subaru executives noted that the sedan version of the Outback accounts for only about 8 percent of the model's sales, or about 3,500 vehicles a year; the rest are wagons. But critics say the actual numbers are less important than the precedent that the reclassification would set.

"If they can do it with a sedan, then anyone can do it with a sedan," said John DeCicco, a senior fellow and fuel economy expert at Environmental Defense. "It's almost like anything goes at this point."

Federal regulations originally set less-stringent fuel economy and emissions requirements for light trucks to avoid penalizing builders, farmers and other working people who relied on pickups. But the exemption opened the way for automakers to replace sedans and station wagons with vehicles that fit the definition of a light truck, notably sport utility vehicles and minivans.

Light trucks now account for more than half of all passenger vehicles sold in the country, up from about a fifth in the late 1970's.

The Transportation Department oversees corporate average fuel economy regulations and fines companies that do not comply with the rules.

Companies that change a borderline vehicle can benefit in two ways, because a big wagon that can sink an automaker's car average may improve its truck average. That, in turn, makes it possible to produce more big trucks and still meet the overall truck standard.

Since the regulatory system was put in place after the oil shocks of the 1970's, the industry has not only invented the minivan and greatly expanded the sport utility and pickup markets, but also started selling wagonlike "crossover" vehicles, like Chrysler's PT Cruiser, that blend cars and S.U.V.'s but are designed to meet the specifications of light trucks.

There are different ways to make a car meet the federal definition of a light truck, including making the rear seats removable to give a wagon a flat loading floor or raising a vehicle's ground clearance to at least 20 centimeters, or a little less than 8 inches. Subaru will raise the Outback's height from a minimum of 7.3 inches to as much as 8.7 inches next year, and will make other adjustments, like altering the position of the rear bumper, to meet light truck specifications.

Significantly raising the ride height can have a hazardous effect on a vehicle's stability. Part of the current Outback's appeal is that it performs better than S.U.V.'s on rollover tests.

"I live in the northern suburbs of New York and I saw a lot of S.U.V.'s on their backs like turtles," said Ralph Schiavone, 46, a consultant who lives in Westchester County, N.Y., explaining why he bought an Outback.

Tim Hurd, a spokesman for the National Highway Traffic Safety Administration, a branch of the Transportation Department, said a vehicle either met the specific technical requirements of being a light truck, or it did not. "They aren't a judgment call," he said.

Added to the complexity of the system is the fact that tailpipe emissions of pollutants are overseen by the Environmental Protection Agency, which has classification rules that do not match those of the Transportation Department. The E.P.A., however, has said it will phase out the distinction between cars and trucks this decade.

Congressional efforts to change fuel economy standards face entrenched opposition from some members of both political parties. But last month, the Bush administration proposed an overhaul of fuel regulations for light trucks and an altered definition to rein in classification problems.

Environmental groups and consumer advocates have generally criticized the administration's proposals as potentially making a complicated system even more prone to manipulation, though they say aspects of the plan—in an early, undetailed form—could be beneficial.

■ Flexibility Urged against Pollution

ERIC PIANIN, WASHINGTON POST STAFF WRITER

Washington Post   January 18, 2001 (Edition F)
Section A
Record Number: 011801XA12Fl6579

Christine Todd Whitman, President-elect Bush's choice to lead the Environmental Protection Agency, signaled yesterday that the new administration will grant state and local governments and industrial polluters more flexibility in meeting federal anti-pollution standards.

While promising to maintain a "strong federal role" in efforts to improve air and water quality and to clean up toxic waste sites, the two-term New Jersey governor asserted that the EPA frequently could achieve more through cooperation than by taking polluters to court or issuing stiff fines.

"We will work to promote effective compliance with environmental standards without weakening our commitment to vigorous enforcement of tough laws and regulations," Whitman told the Senate Environment and Public Works Committee at her confirmation hearing. "We will offer the carrot first, but we will not retire the stick." A moderate Republican best known nationally for her tax-cutting policies, Whitman, 54, earned her environmental credentials mainly in the areas of conservation and combating ocean dumping and beach closings. As governor for nearly eight years, Whitman sought to ease the onus of environmental regulations on factories and businesses.

Some environmental groups complained that she slashed the environmental protection budget early in her first term and relaxed enforcement of pollution regulations in favor of voluntary compliance.

Her nomination won plaudits from Republicans and Democrats on the committee and is expected to easily pass the Senate next week.

However, environmentalists and congressional Democrats worry that Whitman may become the Bush administration's agent for rolling back many of tough regulations issued by the EPA in recent months, such as requiring trucks to reduce tailpipe pollution and refiners to make cleaner diesel fuel, and tightening pollution requirements on factory farms.

Whitman carefully skirted questions about where she stood on those regulations, saying they will be thoroughly reviewed after Bush takes office.

Sen. Hillary Rodham Clinton (D-N.Y.), a member of the committee, told reporters later that "I'll be very concerned if there's any attempt to undermine the environmental [rules] instead of building on them."

Sen. Harry M. Reid (D-Nev.), the temporary chairman of the committee, and Sen. Jon Corzine (D-N.J.) also indicated they doubted the efficacy of the EPA routinely seeking voluntary compliance from states or industry.

"Voluntary compliance sounds good, but my experience is that it doesn't work," Reid told reporters after the hearing.

■ U.S. Walks Away from Treaty on Greenhouse Gases

White House Wants Developing Nations to
Meet Standards as Well

EDITORIAL

THE ASSOCIATED PRESS

St. Louis Post-Dispatch (Missouri)   March 29, 2001 (Three Star Edition)
Section: News

The White House said Wednesday that President George W. Bush would
not put into effect the climate treaty negotiated in Kyoto, Japan, but would
seek an alternative that would "include the world" in the effort to reduce
pollution.

Ari Fleischer, the White House press secretary, told reporters that Bush
wanted to work with U.S. allies on a plan that would require developing
nations to meet certain standards.

"To exempt most of the world is not a treaty the president thinks is in
the interest of this country, or would get the job done," Fleischer said. "It's
important to include the world in the treaty, not exempt most of the world."

Democratic leaders in Congress and environmental groups promised to
fight Bush on the Kyoto climate treaty and other recent policies they called
setbacks to the environment.

"This is yet one other case where the president is saying, 'We want a
foreign and environmental policy that is unilateral,' " said House
Minority Leader Richard Gephardt, D-Mo. "This is a go-it-alone
American policy on the environment, and it must not stand. We have to
turn it around."

William H. Meadows, president of The Wilderness Society, said the new
administration had declared "open season on basic environmental laws and
safeguards that protect the air we breathe, the water we drink, the special
places that we love."

Environmental Protection Agency Administrator Christine Whitman told reporters Tuesday that the Bush administration was not interested in implementing the treaty because the Senate was unlikely to ratify it.

She said the administration would remain "engaged" in international negotiations on ways to address climate change. But it was unclear what position the administration intends to take to the next U.N. meeting on the Kyoto accords, scheduled for this summer.

Whitman repeatedly noted that the Senate had voted 95–0 against the United States taking any action on climate change unless developing countries also take some measures to reduce heat-trapping "greenhouse" gases, which are mainly carbon dioxide from burning fossil fuels.

At the White House for a budget meeting, Republican congressional leaders stood behind the administration decision. "The Senate has already spoken on the Kyoto treaty, and we had 95 votes that said we wouldn't ratify the Kyoto treaty because it didn't apply to all countries and it was imbalanced and it set targets that were probably not reachable," said Sen. Don Nickles, R-Okla., the assistant Senate majority leader.

The Kyoto Protocol calls for industrial nations to reduce emissions—at least for the time being. The United States would be required to cut emissions by about a third by 2012.

On a number of occasions, Bush has expressed his opposition to the Kyoto accord, which the Clinton administration had viewed as essential to dealing with the risks of climate change.

Whitman noted that no other industrial country had ratified the agreement. Three weeks ago, Whitman urged Bush to continue to recognize global warming as a serious concern, arguing that to back away from the issue would be damaging both domestically and internationally.

ASCII IMPERIALISM

Daniel Pargman and Jacob Palme

■ Poor Hörby. On the Internet, the municipality of Hörby in southern Sweden is known as Horby (www.horby.se). One part of the name, *-by*, means village or hamlet in Swedish. Unfortunately, the meaning of the Swedish word *hor* is adultery or fornication,[1] and who can pride him- or herself with living in the "village of fornication" or "adulteryville"? Poor Habo. When the municipality of Habo aimed for the Internet, it discovered that its "natural habitat" on the Internet (www.habo.se) was unavailable. The Web address was reserved for municipality use but had, without ill intent, already been picked up by another, quicker municipality, Håbo. What options do Habo *and* Håbo have to carve out unique niches for themselves on the Internet today and in the future?

How could things go so wrong? To explain why Hörby became Horby and why Habo and Håbo had to fight for their presence on the Internet, this chapter takes a closer look at computer character set standards and their consequences. We thereby look at questions whose answers are unknown to the majority of people who use the Internet. In fact, it might be the case that not only the answers but also the questions themselves are difficult to perceive because they have to do with the, in many respects, invisible infrastructure of the Internet. A seemingly simple question for a Swedish computer user such as "Why can't I use the last three letters in the Swedish alphabet (å, ä, and ö) on the Internet?" leads to the much larger question: "Why do people using languages other than English so often have problems using the Internet?" The short answer is that the rules, or infrastructure, that are in place allow for certain things but not for others. A longer and more thorough answer must delve into a whole set of intriguing questions such as: What are the rules (technical

---

1. *Hor* is etymologically the same word as the English noun and verb *whore*.

standards) that determine how languages work on the Internet? How come things are the way they are and not some other way? How and by whom have those rules been chosen? What are the historical, technical, social and economic aspects behind the answers to the previous two questions? What trade-offs between those aspects are taken into consideration when decisions about future standards are made? What are the alternatives (if any)? Could other rules, other standards, work better? These are the questions that this chapter attempts to answer. We, furthermore, do not settle for a dry description of how things are, what they were, and what they are becoming. We also deconstruct the subject of technical standards for communication on the Internet so as to open the subject up to a discussion also of normative issues of what may, what could, and what should.

This is a political endeavor, and we thread our way carefully so as to not step on too many toes. There are, however, a number of value judgments that can be made after reading this chapter. Whose interests have guided the development of standards for communication on the Internet? Have the present solutions solved the problems in the best possible way for:

- ordinary nonexpert users of computers and the Internet?
- software developers and computer experts?
- the reliable operation of computers and the Internet?

We suggest that there is a built-in bias in the current Internet infrastructure. We, furthermore, argue that the interests of English speakers in general and of software developers in particular have much more influence than the interests of nonexpert users and people using languages other than English.

We clarify here the issues and the reasons for and against specific solutions, but in the end refrain from making any strong final value judgment. That decision is instead left to each individual reader.

> While technical decisions are often defended as rational, logical, or pragmatic, the construction of technological apparatus is the result of a complex interwoven process that include both social and technical factors. In this case, "construction" refers not just to the actual coding decisions and program-writing by which software programs are made, but also to the rhetorical and argumentative strategies that "construct" some software programs [or technical standards] as "better" or as "more effective". Ratto (2001)

Beyond the Internet: The World at Large

The background presented next is divided into two parts. The first part deals with general questions of language and standardization. The second part deals with specific question of how the characters that you and I can use when we communicate through computers have been standardized throughout the last decades.

Demographics and Languages

It is often taken for granted that English is and always will be the lingua franca of the Internet and perhaps also of the world at large. This is, however, not the case. Although English with its slightly less than 400 million native speakers is the second largest language today (Chinese is the first and has three times as many native speakers), it is estimated (Wallraff 2000) that English will become one of four languages with around 520 million native speakers in 2050 (520 million ± 40 million—the other three languages are Hindi/Urdu, Spanish, and Arabic). In fact, because English speakers on a global basis have relatively low birth rates, the global proportion of native English speakers is expected to shrink from over 8 percent in 1950 to less than 5 percent in 2050. These projections are relatively sturdy because most of the people who will be around in 2050 are either already around now or will be the children of people who are around now. Any future gains for the English language must therefore come as a second language or a foreign language. Such numbers are more difficult to estimate, and opinions about them vary widely. Estimations of the number of second-language speakers of English vary between 98 and 518 million (Wallraff 2000).

As to English as a foreign language, it is even more difficult to determine who can and who cannot be considered to be an English speaker, although we do know that people regularly rate their ability to speak English higher than it turns out to be when tested. With satellite television, less than half the expected European audiences turned out to be fit for English-language television and less than 3 percent had an excellent command of English in countries such as France, Spain, and Italy (Parker 1995). Only in small countries (and thus small markets) such as Scandinavia and Holland did the numbers exceed even 10 percent. We do know from satellite television—an area more mature than the Internet—that viewers want television programming to be in their local language (Parker 1995). As Internet access continues to diffuse, there is no reason to believe the situation will be different regarding which language Internet users want their Web pages and their e-mail conversations to be in. These questions, in fact, have to do with the very identity of many millions or even billions of people. The ability to express ourselves in our native languages is "the most immediate and universal symbol of that identity" (Crystal 1998, 115).

Taking this into account, it is interesting to review the results of three surveys that WorldLingo (www.worldlingo.com), a company that specializes in translation and localization services, did. In November 2000, April 2001, and August 2001 WorldLingo sent e-mail messages to between 220 and 250 of the largest corporations in the world (WorldLingo 2000, 2001a, 2001b). Each company got one e-mail message with a question written in one of six languages: Spanish, German, French, Italian, Portuguese, and Japanese (split evenly among the languages). Although the surveys were quite small, the target companies were chosen exclusively based on their size (culled from Fortune 500 and Forbes International 800) and the results were telling.

More than half of the companies (64 percent) did not answer the e-mail at all, and in all three surveys less than 10 percent of the corporations answered the question correctly (i.e., giving a correct answer and doing so in the same language the question was posed in). That means that more than 90 percent of the largest companies in the world failed to answer foreign-language e-mail messages adequately. As an aside, when the response rate was broken down by industry, Industrial Components and Farm Equipment had the best response rate (29 percent) and Entertainment and Media the worst (0 percent).

Of the languages used in the survey, Japanese fared the worst. Of more than one hundred messages sent over the course of the three surveys, only 2.0 percent were answered correctly and "many [companies] did not realize they were receiving a foreign language e-mail" and thus "replied in English saying the email was corrupted" (WorldLingo 2001b). WorldLingo suggests that most companies did not even realize that they were receiving e-mail written in Japanese because their e-mail clients could not handle Japanese characters.

A company such as WorldLingo naturally has a vested interest in being able to present figures such as these, but that does not make the results any less interesting. The conclusions of WorldLingo are also supported by other (market) research from a few years back that showed that more than 50 percent of the Internet users speak a native language other than English (Global Reach), that 37 million Americans do not speak English at home (U.S. Department of Health), that Web users are up to four times more likely to purchase from a site that communicates in the customer's native language (IDC), and so on.

Let us widen the WorldLingo question about whether companies can answer e-mail in half a dozen world languages. Another question well worth posing is: What are the effects of English and other languages on the approximately 6,500 different languages spoken on Earth today? It is estimated that an average of two languages disappear every month (Schauer 2003), and Michael Krauss (1992) estimates that 90 percent of all languages may die out within this century. Thomas Schauer bleakly notes that "it can be expected that in the long run only those population groups with sufficient financial resources to be active customers will be addressed in their native languages [on the Web]. However, many cultural minorities are disadvantaged financially and cannot provide a market which would be large enough to make websites in their language profitable" (2003).

To recapitulate, almost two-thirds of the companies in the WorldLingo surveys did not answer a request in a foreign language, and those that did often asked for the e-mail to be re-sent in English. We are not primarily interested here in whether or how companies answer e-mail questions posed in different languages. Our main interest is, rather, in the more basic question of whether (non-English) users have adequate possibilities to write messages in their own native languages in the first place.

Standardization

Over the last centuries, as trade has increased, distances have shrunk, and countries have gotten increasingly tied up in the business of other countries, standards have become increasingly important for the smooth functioning of the modern world. Modern technology needs standards so as to allow products made by different manufacturers to work together. A simple case is a standard for nuts and bolts that allows us to easily buy a bolt that fits into a threaded hole. "Standards are absolutely necessary for the [information infrastructure] to exist. To be able to communicate, partners have to use a common standard—that is, a 'language,' or, in more technical terms, a protocol" (Hanseth, Monteiro, and Hatling 1996, 410). In the area of computer communication, standards aim at allowing any Web browser to show any Web page (even if the Web browser and the Web page—authoring software were made by different vendors) and for any e-mail to be successfully handled by the e-mail software used for writing, transporting, and receiving/displaying the message (even if the software programs were made by different vendors).

Bowker and Star note that "Information infrastructure is a tricky thing to analyze. Good, usable systems disappear almost by definition. The easier they are to use, the harder they are to see" (1999, 33). Although information infrastructure standards easily slip into invisibility, the modern "made" world is at the same time brimming with standards:

> Classification schemes and standards literally saturate our environment. In the built world we inhabit, thousands and thousands of standards are used everywhere, from setting up the plumbing in a house to assembling a car engine to transferring a file from one computer to another. Consider the canonically simple act of writing a letter longhand, putting it in an envelope, and mailing it. There are standards for paper size, the distance between lines in lined paper, envelope size, the glue on the envelope, the size of stamps, their glue, the ink in a pen, the sharpness of its nib, the composition of the paper (which in turn can be broken down to the nature of the watermark, if any; the degree of recycled material used in its production, the definition of what counts as recycling), and so forth. (Bowker and Star 1999, 37)

We are not primarily interested here in the standardization of time and space, or of weights or currencies, but in the trend toward the standardization of *language* on the Internet and its intersection with technical standards for communication on Internet. What can and what cannot be expressed when it comes to electronic communication is, in the end, determined by the underlying and in many respects invisible infrastructure of standards that enables (and, at the same time, constrains and restricts) such communication.

If the information infrastructure both is ubiquitous and tends to disappear and become invisible, how can we get hold of and get a good look at it? Geoffrey

Bowker and Susan Leigh Star (1999) suggest four strategies for "infrastructural inversion" (Bowker 1994b) to enact a gestalt switch and make the invisible visible. One strategy for understanding a phenomenon is to study its history and development over time (its genesis). So, in the second part of the background given here, we present a short introduction to the history of standards for the most basic building blocks of all electronic communication—the very characters that constitute the domain names (uniform resource locators, URLs), the e-mail addresses, and the actual messages that we exchange on the Internet.

On the Internet: Character Set Standards

We provide here short descriptions of four important character set standards[2] as well as a standard for switching between different character sets in the middle of a text. These standards are ISO 646 [ASCII], ISO 8859, Unicode, ISO 10646, and the ISO 2022 switching standard.[3] Note that character set standards have been around for longer than the Internet, but for the purpose of this chapter we are interested only in their effects for (interpersonal) communication—something that has exploded during the last decade.

It is very difficult to specify exactly when a standard is deployed because standards do not spring forth into the world fully formed. Standards are instead discussed, negotiated, and hammered out over time. They go from being "under development" to becoming "proposed standards" or "draft standards" to finally becoming just "standards" (or "standards-in-use" or "full Internet Standards"). Despite this, a draft or proposed standard is much more finished than the name implies; it is, in fact, the draft/proposed standard that is the most important version because it is this version that is implemented. Everything that happens thereafter are minor corrections as the standard is adapted, corrected, and slowly stabilized over time. Furthermore, even if it is possible to track down the exact date of when a draft standard was proposed, that does not say much about when the standard was actually accepted and deployed because this is done at different times by different relevant actors. It follows from this that it is quite difficult to precisely date a standard.

ISO 646

ISO 646 was the dominant character set in the 1970s. It contains ninety-five printable characters, and each character is represented by 7 bits (each bit represent-

2. There are also a number of alternative standards (e.g., EBCDIC and ISO 6939) that are not relevant for the argument made in this chapter.
3. There are also a number of other technical standards that relate to the subject at hand and that, for example, restrict or put special requirements on which characters are allowed in an e-mail address [RFC1522/2047], on which characters are allowed in a domain name/URL [RFC3490/RRC1738], on which characters are allowed on Web pages [HTML 4.01], and so on. We do not describe these technical standards here per se, but we come across some of their effects later in the chapter.

ing a 0 or a 1) in the computer.[4] ISO 646 replaced various earlier (national) 6-bit character sets that allowed for only UPPERCASE LETTERS. The ninety-five printable characters of ISO 646 include the Latin (English) alphabet, Arabic numerals, and common punctuation characters. Certain characters in ISO 646 are marked for national use, and ASCII (American Standard Code for Information Interchange) is in fact not a separate character set but, rather, the U.S. variant of ISO 646. The slot that ASCII uses for the character "{" is, for example, the one used for the character "ä" in the German variant of ISO 646. It is possible to switch between different ISO 646 variants by using switching codes (see the section "ISO 2022 Switching Standard").

ISO 8859

With the increased international use of computers and computer communication, ISO 8859 was deployed in the 1980s (predating the international dissemination of the Internet). Each character in ISO 8859 is represented by 8 bits (called an octet) in the computer and ISO 8859 contains 190 printable characters. The Western European variant, ISO 8859-1, makes allowances for important characters from Western European languages other than English such as the German (and Swedish) "ö," the French "é," and the Spanish "¿." Again, there are different variants of the ISO 8859, such as the central and Eastern European variant (ISO 8859-2), the Cyrillic variant (ISO 8859-5), the Greek variant (ISO 8859-7), the Hebrew variant (ISO 8859-8), the Thai variant (ISO 8859-11), and so on. Again, it is possible to switch between different variants by using switching codes (see next section). In character sets with more than 190 characters, such as the Japanese Kanji, each character is represented by two octets in a sequence.

ISO 2022 Switching Standard

ISO 2022 is a standard that allows a text to include switching codes. These make it possible to switch to and fro between different character sets (languages) within one and the same text. ISO 2022 can, in principle, be used to switch between any two variants of the ISO 8859 character sets, but in practice only certain much-used combinations are supported by the available software. ISO 2022 is much used in Japan as a means of mixing Japanese and English, and a text can thus begin in Japanese, switch to English, switch back to Japanese, and so on.

Unicode

Unicode was developed during the 1990s[5] to solve many of the problems that appeared with several parallel variants of previous characters sets. Unicode origi-

---

4. That each character is represented by 7 bits means that there are 2^7 (i.e., 128) ways to combine the seven 0s and 1s. Of these 128 ways, 95 are allotted to printable characters, and the others are used to represent formatting (e.g., moving the print position, line breaks, page breaks), ISO 2022 switching codes, and so on.

5. The Unicode character set standard is still being developed today.

nally allowed for around 65,500 printable characters; each character is represented by 16 bits (two octets). Unicode has recently been extended so that in some cases it uses three octets, thus further increasing the number of allowed printable characters.

ISO 10646

ISO 10646 has been proposed as a future standard and is designed to allow for a whooping 4.295 billion different possible characters (4,294,967,296). The intention is that it should be able to harbor every character found in every language on Earth. Each character in ISO 10646 is represented by 32 bits (four octets) in the computer.

At present, a subset of ISO 10646 is used that is practically identical to the Unicode standard. There are only two things that differ between these two standards, and one of them is their names. The only other difference is that ISO 10646 is designed from the beginning to allow for more future expansion than Unicode. Unicode and ISO 10646 can, therefore, in practice be regarded as two identical character sets at the present time. Unicode/ISO 10646 is today supported by the vast majority of hardware and software companies.

Current Situation

The Internet started out as a U.S. network that used only ASCII. Different standards organizations such as the Internet Engineering Task Force (IETF) and World Wide Web Consortium (W3C)[6] have most commonly met the problem of catering to other alphabets by mapping "new" or "nonstandard ASCII" characters on to some more or less arbitrary combinations of "old" ASCII characters. Table 7.1 shows how the word *Hörby* is mapped according to various standards.

As is shown in table 7.1, the encoded text is not easy to read for a human. Fortunately, no human is ever supposed to see the encoded text; the codings in the table illustrate only how *Hörby* is coded and stored inside computers. To repeat: the different varieties of encoding in table 7.1 should always be hidden from the user by the software, and nothing but *Hörby* should, at least in theory, ever be displayed on a computer screen to ordinary computer users.

The advantage of this method of mapping non-ASCII characters on to some sequence of ASCII characters is that it is backward compatible so an old mail program will not suddenly and unexpectedly terminate (crash) when it receives text that is encoded H=?iso-8859-1?Q?=F6?=rby. Such a program would otherwise be at risk of crashing whenever it encountered unknown character codes

6. IETF is the most important standards organization for the Internet. W3C is the most important standards organization for the World Wide Web.

Table 7.1. Varieties of encoding

Encoding of the word Hörby	When it appears in the following places (and according to the following standards)
H=?iso-8859-1?Q?=F6?=rby	E-mail subject header [RRC1522/2047]
H=F6rby	E-mail text body[a] [RFC1521/2045]
Handouml;rby	HTML code [HTML 4.01]
xn—hrby-5qa	Internationalized Domain Name standard [RFC3490]

[a] This (e.g., "quoted-printable encoding") is actually only one of several possible ways to encode the letter "ö" inside an e-mail message.

(e.g., the ö in Hörby) pertaining to since-then-developed standards.[7] But even though the relevant standards in many cases have been in place for years and years, and sometimes even a decade or more, mail programs do not always work the way they are supposed to. When this happens, a program that does not understand what to do with special encodings (as in table 7.1) will instead display a text that contains said codings, which is very unnatural and has a high gibberish factor for the user.

Ideally this should never happen or at least happen only to old programs during a transition period; but in practice it happens quite often. The planned-for case—perceived as a necessary evil—happens when an old program displays this type of gibberish. The unplanned-for case also happens when new programs are designed without taking established standards into account—thus forever (or so it seems) condemning non-English speakers to settle for gibberish-infested language.[8] The standards for e-mail headers [RFC1522/2047], for example, are by now more than a decade old, but there are still many software products that do not correctly support it and that thus display the internal, difficult-to-read gibberish encoding to ordinary users. Many other software products work correctly for the simplest cases but not for special cases. For example, one of these special cases is copying text from a previously received e-mail message into a new e-mail message. This is a very common action; at the same time, it is handled incorrectly (e.g., not according to the relevant standards) by many software products. Because the problems described here happen only to special (non-ASCII) characters, an important characteristic of

7. As a side issue, note that a well-written program should, of course, *never* crash no matter what! Even though an old well-written program (for natural reasons) will not be able to correctly handle later written standards, it should still be able to handle such (and other) input without crashing. In addition, people developing standards have as a goal that standards should work, if possible, also for not-so-well-written programs.

8. Note that those who made the standards do not agree with this. They will say that, even though the standard is patched on top of English, this should not be any problem to users because correct implementations of the standards never show the confusing coding, only the coding unpatched into correct text in the character set used.

the current solution is that, no matter what adverse effects it may have in different parts of the world, it never has any adverse effects at all on texts written in English. Such texts will be displayed correctly by virtually every mail program ever created. The problems described here are thus rendered all but invisible and undetectable to monolingual English-speaking users.

Alternative Solutions

So far we have pointed out problems with the current solution to the language problem in computer-mediated communication on the Internet. Unfortunately, each possible alternative solution comes with its own set of advantages and drawbacks. We here briefly sum up a variety of possible solutions so as to make them easier to compare (see table 7.2). The first two examples are not realistic solutions (ASCII rules and Tower of Babel) but are, rather, included to illuminate extreme positions better.

It is possible to implement each of the proposed solutions, in principle, because any resulting technical challenges could feasibly be solved by throwing enough resources into solving them (i.e., compare with the solution to the Year 2000, Y2K, problem[9]). The choice of which solution is the best one is ultimately not something that can be determined by counting, weighing, and measuring the relative merits of the proposed solutions but, rather, is a matter of making difficult trade-offs among different (legitimate) technical and social choices. Standardization is only the *process* of negotiating, deciding, and enforcing one solution rather than another, but the process of standardization itself has very little to say about which values are more important than others, beyond the simple axiom that any solution that can be agreed on is better than having different parallel and competing solutions.

It is easy in a chapter such as this to allow ourselves to think freely beyond the mundane reality of how things are, what they were, and what they will become. In the real world—a place where things should work tolerably (rather than perfectly) and, preferably, yesterday—things are not that easy. In the real world, we are restricted by decisions that were made long ago and by the resulting inertia that has accumulated over the decades. The most common solution is to patch things and hobble along. Any deviation from that solution requires potentially large alterations of the existing information infrastructure, and such alterations are *expensive*. To patch things, instead, is less costly than other, more substantial changes.

9. The Y2K problem stemmed from decades-old computer programs' not being adequately prepared for the turn of the twenty-first century. There was widespread concern that crucial systems would cease to operate at midnight, December 31, 1999, and it has been estimated that more than US$300 billion were spent on preparations during the preceding years.

Table 7.2. Comparison of methods when adding international character sets to standards

Name	Description	Technical advantages/disadvantages	Social advantages/disadvantages
ASCII rules	A step back in time that forces everyone to always use ASCII everywhere.	Large technical advantages because it forces everyone to use the same restricted set of Latin/ English characters.	Large social disadvantages because it forces everyone to use the same restricted set of Latin/ English characters.
Tower of Babel	A possible reaction to an ASCII rules policy; different local character sets are invented in the absence of a common technical standard that can satisfy a majority of Internet users.	Large technical disadvantages because individual "bridges" have to be constructed between each pair of different character sets/languages. Goes against the direction of changes in the modern world (increased standardization, rationalization, communication, etc.).	Large social advantages for smooth communication between people using the same character sets/languages. Large social disadvantages because communication between people in different countries becomes difficult or impossible.
Patch it up	Encodes or translates new character sets into obscure sequences of characters in older sets.	Old software that does not comply with the new standard will seldom crash. Relatively inexpensive to implement. "Glosses over" older technical design decisions that might not have been good in the first place. Introduces layers of complexity that cause unexpected difficult-to-diagnose problems when special cases inevitably crop up. Each special character is replaced by seveal other characters that can cause	No obvious disadvantage when all software supports the standards perfectly. Old software does not support new standards, and non-English speakers will see encoded text that is obscure and confusing. In reality, software is often not correctly written; also new software that does not support the standard will at times display obscure and confusing encoded or text.

(continued)

Table 7.2. (*continued*)

Name	Description	Technical advantages/disadvantages	Social advantages/disadvantages
Leave the past behind	Make a clean break and change to a new and improved character set standard, such as Unicode or ISO 10646.	problems when there are limitations on length (e.g., in e-mail headers). Easy-to-write, new, good software when the past can be left behind.	Best possible solution with large social advantages when all software support the standard perfectly.
		Large technical disadvantages because old software may crash when it receives text with the new standard.	Technical disadvantages turn into social disadvantages because old software crashes and new software has to be purchased/installed.
		Large technical disadvantages because it forces everyone to rewrite all existing software.	
Invisible plurality	Use a standard such as ISO 2022 to switch between different character sets in a text.	Neat technical solution, but very complex and difficult to implement.	Social advantages when all software support the standard perfectly.
		Not fully explored technically; today it is well supported only in special cases, such as for combining Japanese with other languages.	
Dual identity	E-mail addresses and domain names can have two different addresses, one local and one English or international.	No technical disadvantage as long as the English name is used (see ASCII rules).	Social advantages because anyone anywhere can always fall back on and use an English or international name.
		Not supported in existing software and therefore not fully explored technically.	Complex to remember two names for the same object and to use the right name in the right place; easy to send a national name to someone outside the national language area.
		Potentially technically complex to handle two names for the same object.	

How Standards Develop

> What appears as universal or indeed standard, is the result of negotiations, organizational processes, and conflict. How do these negotiations take place? Who determines the final outcome in preparing a formal classification? . . . Someone, somewhere, must decide and argue over the minutiae of classifying and standardizing. . . . Once a system is in place, the practical politics of these decisions are often forgotten, literally buried in archives (when records are kept at all) or built into software or the sizes and compositions of things. (Bowker and Star 1999, 44–45)
>
> [. . .] many layers of technology accrue and expand over space and time. Systems of classification (and of standardization) form a juncture of social organization, moral order, and layers of technical integration. Each subsystem inherits, increasingly as it scales up, the inertia of the installed base of systems that has come before. (33)

On a practical level, many standards for computer communication are developed by standards organizations such as the International Standards Organisation (ISO), International Telecommunications Union (ITU), IETF, and W3C. Some organizations are more bureaucratic, whereas others instead rely on a rough consensus for making decisions. User influence is usually very limited because the actual work is done mostly by technical experts within these technocratic/meritocratic organizations. Most experts represent companies that are active in developing computer applications. The U.S. influence is usually very strong—for example, in the IETF around two of every three people are from the United States. We use e-mail as an example here to show how standards are developed and to point out what issues are taken into consideration in that process.

Even if standards organizations managed to create new standards (e.g., new character set standards) that were fair to all, software programs, such as those for e-mail, are built by commercial actors rather than by the standards organizations. Commercial actors can choose to follow the current and emerging standards—or not. If the standards organizations were to arbitrarily impose new standards and then were able to enforce these standards, it would create large problems for many existing e-mail programs. This constitutes a strong argument for making a few changes slowly so as not to make a clean break with the past and against making faster or more radical changes that "once and for all" solve the problems the right way (whatever that is) and amend past injustices.

The attitude in the standards organizations is, therefore, geared toward solving a few select technical problems at a time while not creating other, new technical problems. More radical changes could, in a worst-case scenario, even have network-wide repercussions and disturb the smooth functioning of parts of or the whole Internet—although no one can know for sure about possible future effects in advance. The prevailing atmosphere in the decision-making process is generally conservative and risk aversive when it comes to taking technical risks. New technical standards for computer applications should *always* be backward compatible; that is, new standards must not make old nonadjusted programs out there cease to work— the mail must get through! *Work* in this context means that the programs should function primarily in a technical sense, by not suddenly and unexpectedly crashing. That the program should *work* in the sense that it should function well for the human beings who use it is thus implicitly deemed less important.

As already mentioned, the U.S. influence in standards organizations is usually very strong, but the U.S. dominance is even stronger when it comes to the market for developing and selling e-mail programs. Most mail programs in use were developed in the United States, and it is generally not easy for American monolingual managers and developers to understand the problems that noncompliance with existing standards can cause for people in other parts of the world.

The standards that are chosen by the standards organizations and commercial actors therefore tend to be conservative and to (1) work very well for the English language, (2) not cause existing programs to crash, and therefore (3) make few demands on commercial actors to rewrite or adapt existing programs. As it so happens, the interests represented by the people who decide on the standards thus correlate to a high degree with the interests of (primarily) U.S. software developers. More generally, these decisions tend to benefit all English-language users to the disadvantage of those who use any other language; they therefore fall within Batya Friedman and Helen Nissenbaum's definition of *bias* in computer systems. Friedman and Nissenbaum use this term "to refer to computer systems that *systematically* and *unfairly discriminate* against certain individuals or groups of individuals in favor of others" (1997, 23; emphasis in the original).

This line of reasoning is, however, not to say that decisions made about standards and how to apply them are necessarily bad per se. There are obvious advantages—for everyone—in having a system with limited functionality that works well enough for everyone (albeit better for some) rather than having a system with extended functionality that works less well—for everyone. To have a system that works well enough for everyone is, after all, the very purpose of agreeing on a standard in the first place. The same reasoning can be applied more generally to using English as a lingua franca both on and off the Internet. It creates a common playing field; even though that playing field is not level and benefits native English speakers the most, it still provides people from different cultures a way to communicate without knowing one another's native language.

For better or for worse, this is not a simple black-and-white tale of good versus evil but a rather more complex situation in which it is difficult to tell what is right and what is wrong and in which complexity and difficult trade-offs are hidden in many places. This chapter describes how standards are created. The question of how standards *should* be created is still an open and much more difficult question. It is sufficient to say is that there is no one best standard because every proposed solution represents a trade-off between different interests. Any proposed, approved, or realized solution solves certain problems but at the same time avoids, creates, or exacerbates other problems. A proposed solution can solve some technical problems while at the same time refraining from solving other technical problems, or it can solve some technical problems while creating *new* social problems for groups of individuals. Any specific solution tends to spread its graces unevenly, solving some problems deemed important for certain groups of stakeholders while failing to address, disregarding, or creating other problems for other stakeholders. Such is the nature of standardization work. The creation of new technical standards is a messy and time-consuming process of negotiation among different conflicting interests, interests that can all be perceived as legitimate depending on our perspective. Which solutions are adopted and which are rejected? The answer to that question is intimately connected to issues of power and representation in the fora where decisions are made.

How Standards Are Implemented

The standards specified in standards document (formal standards) and the standards used in real practice (de facto standards) are two different things. In real practice, usually only a subset of a standard is implemented, but at other times an extension is used that goes beyond the formal standard. Software vendors tend to implement only the part of a standard that is necessary for their software to work. In practice, this means that vendors tend to support what other vendors support and *not* everything that is formally specified in standards documents. This difference can sometimes be very large. In particular, mixing languages (see the section "ISO 2022 Switching Standard") is often not supported adequately because vendors do not see a user need for switching between two languages within the same text.

Some software standards, for example those for e-mail, have many hundreds of different implementations, made by different companies. All these implementations have to work together with all the others, and this makes progress difficult. If a vendor adds a new feature that extends or requires changes in the standard, that feature can at first only be used with other software from the same vendor. Even when an agreement has been reached about the extension of a standard, some standards still never get widely adopted by the market. There are several examples of this, such as the existing standard for nondelivery notification for e-mail. It is also theoretically possible according to existing standards for an e-mail program to allow people to use an e-mail address that contains blank spaces (e.g., between the given name and family name) except for the not irrelevant fact that a large number of existing e-mail programs are not able to handle the coded address correctly.

Standards organizations do not unilaterally change and heavy-handedly enforce new standards, and commercial actors can choose not only to not adopt a proposed standard but can even choose to develop their own (competing) standard.

The result is that any standard that has been developed by a standards organization *or* a commercial actor can end up becoming widespread or not used at all (or anywhere in between). The larger the software company, the larger the chance of succeeding in getting other actors to accept and adopt a proposed standard.

Large and rich companies often try to influence standards-making to produce large and complex standards because such standards are more effective in excluding competitors that cannot afford to implement them. When Internet standards were made mainly by academic people, they were usually simple and easy to implement; the influx of people from large commercial companies is one reason why standards today generally tend to become more and more complex. One of the more devious ways for a company to create a business advantage is to de-commoditize a commodity protocol (standard) (Levien 1998). This strategy has also been called "embrace and extend," but both terms are in reality euphemisms for hi-jacking or privatizing a protocol by forcibly extending it so that it does not work with the old version of the standard anymore. "In an ideal world, software would be created to fulfill user needs, and would be designed to be maximally usable. However, real software gets created for somewhat different aims, sometimes in line with these user needs, sometimes less so. Proprietary software is created to make money for its authors" (Levien 1998).

A very good way to make money is to own a standard. Standards organizations actively resist the de-commoditization of protocols by working to develop commodity protocols (level playing fields that benefit all actors rather than some specific actors). But really profitable software manages to create a proprietary advantage by raising barriers to competition. There are different ways to take a commodity protocol (standard) and de-commoditize it and Ralph Levien (1998) enumerates six such ways (e.g., specify the "improved" version incompletely and change the standard rapidly or make it more complex), of which only one actually benefits the end users by adding value through improving the protocol. The larger the company and the better and more useful the software product/protocol/standard, the larger the company's chance of succeeding in getting other actors to accept and adopt it as a proposed standard for reasons of interoperability.

Analysis

As we have emphasized in this chapter, any proposed solution to the language problem on the Internet has technical, social, economic, and cultural consequences. This is naturally also true for any non-solution (e.g., continuing on the beaten track without much contemplation or discussion of the problem). Social, economic, and cultural issues often go together (e.g., if a mail program cannot accommodate certain users, it will not be economically successful in that market) and have for the sake of simplicity been lumped together as "social" factors in this chapter. This makes it easier to differentiate such softer issues from and contrast them with the harder technical issues. Any deeper discussion of the subject would, however, benefit from a decomposition and an analysis of the specific aspects of said social factors.

A critique of any specific standard(s) represents only one side of the issue. We next present two different kinds of analysis that go beyond the simplistic question: "English—is it good or is it bad?" The first takes as its starting point an article by Friedman and Nissenbaum (1997) and concerns what bias-free software could mean in this context and how it should be designed. The second takes as its starting point a sleek book by Pierre Guillet de Monthoux (1981) and critically examines the practice of constructing standards and the potentially negative effects of inaugurating a multitude of new standards.

Bias in Character Set Standards

In their article "Bias in Computer Systems," Friedman and Nissenbaum (1997) start by defining *bias* in computer systems (i.e., "computer systems that *systematically* and *unfairly discriminate* against certain individuals or groups of individuals in favor of others") and then describe how bias can be built into computer systems. Just as the code of a computer program is hidden inside a black box (i.e., not be readily discernable to the majority of people who use it), built-in bias often becomes difficult or impossible to remedy or even identify. Friedman and Nissenbaum set out to define a framework with different categories of bias in computer systems and then go on to use that framework to analyze three biased computer systems. Finally, they wrap up their article with suggestions for minimizing bias in computer system design—a theme that has been further developed in Friedman's (1996) article "Value-Sensitive Design."[10] Their framework for analyzing bias and their suggestions for the design of less-biased or unbiased computer systems are of special interest here. How can the design and application of character sets and, more generally, the infrastructure for communication over the Internet be analyzed in terms of built-in biases?

Categories of Bias   Friedman and Nissenbaum differentiate three broad categories of bias in computer systems: preexisting bias, technical bias, and emergent bias. Preexisting bias "has its roots in social institutions, practices and attitudes" (explicit or implicit, conscious or unconscious), and technical bias "arises from technical constraints or technical considerations" at the time the system is designed. Emergent bias—in contrast to the other two types of bias—is impossible to identify at the time of the creation or implementation of the system and, instead, typically arises over time, in the context of using the system. Emergent bias typically

---

10. The articles are published in the "wrong order," but aspects of the work on bias had been presented at conferences five years before "Bias in Computer Systems" was finally published in 1997.

emerges as a result of changing societal knowledge, changing population of users, or changing cultural values.

Using Friedman and Nissenbaum's framework, we can see that character set standards suffer from all three types of bias. The original bias was technical and arose due to technical constraints. With the different variants of ISO 646 limited to ninety-five printable characters (in the 1970s), some tough choices had to be made as to which (limited) set of specific characters to include. Despite the later replacement of ISO 646 by other standards, it is only with the possible full and complete future adoption of the ISO 10646 character set standard that this technical-constraints bias will be totally eliminated.

The second type of technical bias is due to technical considerations, especially considerations of the backward compatibility of software. It is difficult to separate this type of technical bias from preexisting bias in the organizations, institutions, and individuals "who have significant input into the design of the system" (Friedman and Nissenbaum 1997, 25). Friedman and Nissenbaum break down preexisting bias into two subcategories, individual and societal preexisting bias; both kinds of bias are present in the decision-making processes that are described here.

The third category of bias, emergent bias, becomes apparent only over time. At a time (some decades ago) when the vast majority of Internet users were English speakers, some types of bias could not have been recognized as bias at all. It is only later, "when the population using the system differs on some significant dimension from the population assumed as users in the design" (Friedman and Nissenbaum 1997, 27) that emergent bias can be identified. The category of emergent bias thus seems to refer only to the temporal dimension of when social or technical bias was discovered rather than to a qualitatively different *kind* of bias—we cannot differ emergent bias from the previous two categories of bias. It is easier for us to think of emergent bias as representing a "latent dysfunctionality" (Merton 1957) or a ticking time bomb that was built into the system from the beginning (albeit only discovered later).

Designing to Minimize Bias   Friedman and Nissenbaum (1997) suggest that bias can be remedied by, first, identifying bias in any given (existing) system and, second, by developing methods for correcting bias in old systems as well as avoiding building bias into new systems. Some methods that they suggest are primarily aimed at developing bias-free application software and are of less use in a discussion of standards for an information infrastructure (e.g., you do not use rapid prototyping or test groups when you develop a character set standard).

We have here taken only the first of these two steps by identifying bias (and the effects of that bias) in the information infrastructure of character set standards. To suggest practical guidelines for developing a bias-free infrastructure is beyond the scope of this chapter. Leaving aside the question of what should be done in the future, what could or should reasonably have been done to develop bias-free character set standards in the past? To apply Friedman and Nissenbaum's guidelines is both a test of past decisions about character set standards and a test of how reasonable the guidelines themselves are in this case.

To minimize preexisting bias, their advice is for designers in the earliest stages of the design process to scrutinize the design specifications. This activity should be coupled with a good understanding of relevant biases out in the world, such as biases based on cultural identity (including language), class, gender, literacy, handedness, and physical disabilities. To minimize technical bias, they propose that the designer think beyond technical features that are internal to the system and envision it in a context of use and "that technical decisions do not run at odds with moral values" (Friedman and Nissenbaum 1997, 37). This sounds reasonable but difficult to do, and no further advice is given for how to achieve this goal.

Giving normative advice for minimizing emergent bias is, unsurprisingly, most difficult. It is hard to posit guidelines for how to design future-proof software so that it minimizes or avoids bias will emerge at a later time. Friedman and Nissenbaum (1997) ask designers to account for increasingly diverse social contexts of use. They admit that questions of what sort and how much diversity should be designed for merit a lengthy discussion, and they settle for offering three suggestions:

1. Designers should "reasonably anticipate" probable contexts of use and design for these.
2. Designers should attempt to articulate constraints on the appropriate contexts of the use of the system.
3. Designers and administrators should "take responsible action if bias emerges with changes in context" (1997, 37).

To what extent have the designers in this case lived up to Friedman and Nissenbaum's advice? They have *not* lived up to these design criteria for bias-free software to a very high extent. Part of the reason for this is that such goals have been subordinate to other, purely technical goals. Still, we feel that it is difficult to hold the original designers in the far past accountable for design decisions that resulted in built-in kinds of bias that only became apparent later. The subsequent success of the Internet and the reach and durability of the ASCII character set must surely have come as a surprise, and their designers in the early 1970s could not have "reasonably anticipated" events that lay one or several decades into the future. It is, however, easier to hold subsequent generations of designers accountable for not doing more than patching up the biased infrastructure for communication over the Internet. What subsequent designers especially have failed to do is to "take responsible action if bias emerges with changes in context." The changes in context (due to the spread of the Internet, e-mail, and the World Wide Web) have at this point been around for more than a decade, and their effects are escalating as more and more non-English speakers appropriate the Internet and integrate it into their everyday lives.

Normative Rationality

Guillet de Monthoux (1981) writes about the intersection of technology, economy, law, and philosophy; he criticizes normative rationality in general and standardization in particular. His main point is that a general drive to standardize everything under the sun ends up having adverse effects on quality. Because the number of technical standards roughly doubled in both Germany and Sweden between 1955 and 1980,[11] following (all) the rules, in his opinion, tends to become more important than making original contributions in any kind of technical practice. The engineer has been transformed from someone who innovates and applies into someone who obeys and enforces (arbitrary but duly decided) standards and rules. In short, Guillet de Monthoux asserts that the dark side of standardization is that the normative, following the rules, tends to replace the qualitative.

Rules encoded in computer code are more stringent than rules (laws and technical standards) written on paper. The software acts as a system of enforcement, similar to a legal system (Palme 1997), but goes further by enforcing obedience and making rule-breaking activities impossible to enact. In short, in computers and on the Internet, code is law (Lessig 1999) in a very literal sense. Although inattention to moral values is disturbing wherever it happens, it might be argued that it is especially disturbing when those values are built into software because we cannot easily negotiate with technology, unlike with people (Friedman 1996).

Guillet de Monthoux's book was written in 1981 and does not concern itself with software, but his analysis can be compared to the present situation of English-speaking dominance in the relevant standardization committees. English-speaking interests will win out over other interests because of their present technocratic, meritocratic, and undemocratic representation on (to the general public) arcane standardization committees. This is especially true when the present dominance is combined with the sunken costs of implementing past decisions made in those committees and with selective references to the importance of following through on some issues (e.g., the backward compatibility of all software) but not on others (e.g., tougher demands on new software to handle newer standards). Guillet de Monthoux's challenge is to ask if—in the process of defining and (selectively) applying standards—the work of standardization committees and software companies actually stands in the way of, rather than supporting, innovation.

Guillet de Monthoux asserts that as standards and the act of complying with them become increasingly important, it also becomes strategically important for companies to (1) take part in the creation of new rules and standards, (2) have inside information about upcoming rules and standards, and (3) follow new rules

---

11. As Guillet de Monthoux (1981) gleefully points out, the German organization for standardization, the Deutsches Institut für Normung (DIN), even has a standard with detailed instructions for how *DIN* should be written.

and standards. The result is, unfortunately, a situation that has certain similarities to an established cartel that excludes small up-and-coming actors that do not have the resources to participate in (international) standardization work. Large companies, furthermore, prefer complex standards that further raise the barriers to entry, and there is thus also a tension between developing simpler versus developing more complex standards.

Levien (1998) argues that simpler standards (also called commodity protocols) will win over complex semi-proprietary and difficult-to-adopt standards in the longer run. Because the latter kind of standards add "gratuitous complexity" (*Levien 1988*), the result is a product that both has higher maintenance costs and becomes harder to adapt as the world and the demands on it change. The simpler standard is, in contrast, characterized by inherent strengths such as flexibility, tailorability, modularity, and openendedness and facilitates ongoing experimentation (Hannemyr 1999).

Guillet de Monthoux (1981) says that participation in standardization committees represents both a technical and an economic threshold that, over time, tends to become more important than the objective qualities of the committee work or the resulting products. The problem is not that a specific company can outright influence the result of a committee to fit that company but that its presence in and of itself (especially when a competitor is absent) becomes the most important reason for participation in the process. Participation in standardization work can thus be seen as a barrier to entry and as a sound investment, and it can lead to an unrestrained multiplication of technical standards as an end in itself. Some companies will take out a concession on a specific and narrow area by establishing a standard in that area. The existence of an established standard then shuts out competing and perhaps better solutions and, by extension, shuts out competitors.[12] Technical and engineering work over time becomes a matter of knowing about protocols and rules rather than applying professional skills to solve concrete problems. The engineer becomes an applier of technical rules or a technical bureaucrat in the service of the normative.

Is this "ASCII Imperialism"?

As we have shown, there does exist a bias among the organizations, institutions, and individuals who have significant input into the design of new standards for computer communication. This bias works in favor of English-speaking Internet users and to the disadvantage of (in varying degrees) speakers of all other languages in the world. This does in itself not constitute information that is in any way new to people who partake in standardization work. Such committees, in general, fully understand all the problems that have been described here, and they still, with open eyes, make the decisions the rest of the world has to live with.

12. See Lehr (1992) for a view that in most respects contradicts Guillet de Monthoux.

This dominance of the English language can be construed as a malevolent imperialistic force that severs people's connections to their (non-English) native languages, as exemplified by a statement made by the director of a Russian Internet service provider: "It is the ultimate act of intellectual colonialism. The product comes from America so we either must adapt to English or stop using it. That is the right of any business. But if you are talking about technology that is supposed to open the world to hundreds of millions of people you are joking. This just makes the world into new sorts of haves and have nots" (quoted in Crystal 1997, 108). Or, at the same time, the dominance of the English language on the Internet can be construed as a lingua franca and a benevolent world standard that provides a common ground for global communication and understanding. We chose the title for this chapter not the least for its potential to provoke and command attention. Although a strong case can indeed be made for the presence of ASCII imperialism, the title does not do enough justice to the positive aspects of having standards (any standard!) for communication—both in terms of languages and in terms of computer protocols. We have refrained here from adopting either of these two simplistic positions, but it does seem reasonable to recommend that, as the context of use changes and more non-English speakers appropriate the Internet, non-English interests be taken into consideration more than has been done up to now.

An obvious objection to this recommendation is that changes to the architecture of an existing infrastructure are impractical and too costly. We object to that argument for two reasons. First, we cannot allow value judgments about the appropriate use of an important international technology to be decided for us purely by what is simple, by what is convenient, and by what is cheap. By doing so, we surrender to what Weizenbaum (1976) aptly calls "the imperialism of instrumental reason." We, the people, are the ones who should decide how technology should work for us and not the other way around. The motto of the 1939 Chicago world fair, "Science Explores, Technology Executes, Man Conforms" is, and should indeed be, a thing of the past!

Second, although the goal of an information infrastructure is by definition to be far-reaching—and thus difficult to change because of inertia—any information infrastructure also needs to be flexible and open to change as the world and the demands on that infrastructure change. Ole Hanseth, Eric Monteiro, and Morten Hatling (1996) make this point when they analyze the inherent tension between the twin demands on information infrastructure: standardization and flexibility. In their article, they compare the inflexibility of the OSI architecture to the (in comparison) nimble architecture of a competing infrastructure for computer networks—the Internet. It would indeed be ironic if the flexible architecture of the Internet has since ceased to be able to change.

We have in this chapter adopted a position that problematizes the decisions of an English-speaking majority to prioritize interoperability, backward compatibility, effectiveness, and convenience over other kinds of qualities when it comes to issues involved in the character set standards for computer communication. An ironic twist is that Swedes are guilty of the very same practices on a local level. For the same reasons of interoperability, backward compatibility, effectiveness, and convenience, we chose to use the ISO 8859-1 Western European version of the 8859 character set standard instead of the specifically Scandinavian version 8859-6. For the vast majority of Swedes (population 9,000,000), the ISO 8859-1 character set standard is "good enough," and adopting it has the added benefit of tying Sweden closer to the global community of speakers of Western European languages (e.g., all other ISO 8859-1 users) and to the existing infrastructure of ISO 8859-1-compliant software. We have thereby sacrificed the possibility of using certain characters that are specific to the Scandinavian languages (e.g., Icelandic, with 300,000 speakers) and the Sami minority language (probably fewer than 30.000 speakers in Sweden and neighboring countries).

■ Bananas, Eurocrats, and Modern Myths

Standards between Power and Agreement

MARTIN LENGWILER

The European Union, notably the Brussels eurocrats, suffers from a bad reputation among its inhabitants. The European Union is notorious for overregulating the most detailed aspects of human life. The Acquis Communautaire, for example, the entire body of European treaties, regulations, and directives, spans more than 80,000 pages and accumulates more than 2,500 new laws every year. Regulations about new standards lie at the core of the EU legislation, particularly after the introduction of the European Single Market in 1992, with which the fifteen member states had to agree on common trade rules. In the tabloids and publications all over Europe, standards and regulations from Brussels are often ridiculed and opposed as a step toward the putative bureaucratic centralism of the European Union (CNN 2008).

Thus, when the news spread in the mid-1990s that the EU administration had defined a standard size of imported bananas, it came as no surprise. The details of the story were complex and opaque, as often is the case with regulations from Brussels. Some newspapers even held that Brussels tried to ban any curved banana from European territories. The British tabloids in particular were happy for another anti-European headline—*The Sun* wrote about the "barmy bunch of Brussels bureaucrats"—and the news made its way into popular lore, where it has flourished ever since (Darnton 1994, A16).

Nevertheless, the better part of the story about the Brussels ban on the curved banana is pure fantasy. The story is one of many euromyths, dozens of fictitious tales about the presumed misdeeds of the EU bureaucracy. Unsurprisingly, this genre of modern myths is particularly popular in eurosceptic Britain. The European Commission Press Office in London has made a virtue of necessity, currently maintaining an excellent and comprehensive collection of euromyths (with ripostes) on its website. The site offers a variety of classics such as the stories about the EU ordering farmers to give toys to pigs, introducing a standard length for condoms, and banning British double-decker buses while introducing the standard "Euro coach" with mandatory seat belts for passengers (European Commission 2008b).

To date, the banana standard remains the archetype of anti-European lore. The origins of the story are misty and contested. The EU office in London argues that it all started with a joke at a press conference in 1992 when the creation of the single market was announced. Making fun of the standardization of trade rules, one British journalist asked what this would mean for the "curvature of bananas." With that ironical question, the banana myth was born. A more plausible explanation refers to a ruling of the European Union in 1994 introducing a minimal size for bananas to be imported to the EU member states. According to existing international standards on the quality and size of traded fruit, the regulation specifies that bananas must be at least 5.5 inches (14 centimeters) long and 1.1 inches (2.8 centimeters) wide. Moreover, they should not be "abnormally" bent. Presumably, the exclusion of abnormally bent bananas was maliciously generalized and reinterpreted as a ban on any curved bananas (Darnton 1994; European Commission 2008a).

The European Union has repeatedly declared that, although bananas of an abnormal shape cannot be imported, this does not mean that the European Union has banned curved bananas, because "a curve is a normal shape for a banana" (Darnton 1994, A16). But the banana myth is still alive today. In 2002, for example, Daniel Hannan, a conservative member of the European parliament from the United Kingdom, sent a written question to the European Commission asking why the Commission tried to deny its unenforceable norms for the curve of bananas. Franz Fischler, the EU commissioner for Agriculture, Rural Development, and Fisheries, called the alleged banana standards a myth and justified European regulations as being in line with similar international trade agreements (Official Journal of the European Union 2003).

Why has the story of the banana standard remained so popular? The explanation has to do with the ambivalent character of standards and the authorities defining them. Standardizing incorporates two processes: consensual agreement among the parties concerned (as the European Union understands it) and the expression of power of an authority over its subjects (as the critics have it). Because of this ambivalent reputation, the setting of seemingly unjustified, or as in this case "unnatural," standards can be criticized as a sign of illegitimacy. Exposing the standards then becomes an effective means of mobilizing opposition against authority, even if the accusation is invented.

Ironically, the European Union does not have much interest in discriminating against nonstandard bananas. In the so-called banana wars that have troubled the trade relationship between the United States and the European Union since 1994, the European Union favored the smaller (i.e., nonstandard) bananas from the Caribbean over the larger and cheaper fruit from Latin American plantations dominated by multinational U.S. companies such as Chiquita and Del Monte. To protect imports from the former colonies of their member states, the European Union set quotas to guarantee the sales of Caribbean bananas in Europe. Pressured by food corporations, the Bill Clinton government filed a complaint against the European Union with the World Trade Organization (WTO) and won the case in 1997. After years of quarrelling, with the United States retaliating with punitive tariffs, the European Union abolished the banana quotas in 2001, a step that brought poverty to the many of the small-scale independent banana farmers in the Caribbean. Against the background of the banana wars, the last thing the European Union wanted to do with its introduction of minimal sizes for bananas was to bar the smaller Caribbean bananas from the European market. On the contrary, the size standards were low enough not to threaten the production from the former colonies (Josling and Taylor 2003).

■ Chocolate Directive Now Agreed

Food Law News—EU—2000
UK Ministry of Agriculture Fisheries and Food
Press Release (82/00)
15 March 2000

After a long period of negotiation, the European Parliament yesterday agreed a proposal for a new Chocolate Directive which will finally create a single market for chocolate whilst recognising the different traditions of making chocolate in Europe.

The main issue on which agreement has now been reached is the way in which manufacturers should be allowed to make small additions of specialised non-cocoa vegetable fats, presently permitted in seven out of the fifteen Member States. After the European Parliament decision in Strasbourg today, all types of chocolate will be freely traded across Europe. For the first time the milkier formulation of British Milk Chocolate will be recognised throughout Europe, although on the continent it will be called "family milk chocolate".

Joyce Quin said: "I am delighted that this important Directive has at last been agreed to the satisfaction of all concerned and that UK interests and concerns have been met. Chocolate is very important not just to the UK but to all European consumers. The Directive permits all traditional and new products to be freely marketed and gives credibility to the Single Market. Consumers have the final say on which products they buy, and we will have a market where manufacturers can offer consumers the widest possible choice of products."

The Parliament adopted the Council's Common Position subject to two minor amendments. Formal adoption of the Directive is subject to final approval by Council. The existing 1973 Chocolate Directive allows UK, Ireland, Denmark, Sweden, Finland, Portugal and Austria to add up to 5 percent of vegetable fat other than cocoa butter to chocolate. It also allows UK and Ireland to sell in their countries a product labelled as "milk chocolate" with higher milk and lower cocoa content than "milk chocolate" made and sold in the rest of the Community.

The new Directive permits chocolate containing up to a maximum of 5% vegetable fats to be freely marketed in each Member State. The addition of vegetable fats required for technical reasons, is in addition to the required cocoa solids (same as in the existing Directive). A separate declaration "contains vegetable fats in addition to cocoa butter". The location of the statement would be in the same field of vision as the ingredients list and near the name of the food. The vegetable fats used are restricted to six tropical plant sources, which reflect the fats already in use.

The Directive would continue to permit UK and Irish "milk chocolate" made to a different recipe to be called "milk chocolate", but, if exported, would have be labelled "family milk chocolate".

Standards Proliferate Even for (Seemingly) Simple Objects

ANSI_ANSLG C78.41-2006
Revision of ANSI C78.41-2001

18-W Low Pressure Sodium Lamp, L69

Bulb: T16 (T51) clear   Base: Bayonet, BY22d

1   Lamp designation

L69/E   (old designation: L69RA-18)

2   Physical characteristics

Nominal diameter..51 mm (2.01 in)
Maximum overall length (MOL)..216 mm (8.50 in)
Light center length (LCL) ....135 mm ± 20 mm (5.31 in ± 0.79 in)
Effective arc length [1] ....85 mm ± 10 mm (3.35 in ± 0.39 in)
Operating position limitation..............................as indicated by lamp manufacturer
Maximum permissible bulb temperature[2].........................150°C (302°F)
Maximum permissible base temperature[3].........................150°C (302°F)
Maximum outline dimensions...see Part III

3   Electrical characteristics

3.1   Lamp operating characteristics. The following lamp operating characteristics apply for lamps that have aged 100 hours. The characteristics shall be measured for lamps operating in a horizontal position in still air at an ambient temperature of 25°C ± 5°C (77°F ± 9°F). The lamps shall be operated on a reference ballast having the characteristics listed in Section 4 with a 60 Hz sinusoidal power supply.

	Min	Nom	Max
Voltage at lamp terminals (V, rms)	47	57	67
Lamp current (A, rms)	–	0.35	–
Lamp wattage (W)	–	18	–

3.2   Warm-up time. The warm-up time shall be measured for lamps operating on a reference ballast at rated input voltage in still air at an ambient temperature of 0°C (32°F).

Maximum time for lamp voltage to stabilize ....15 minutes

---

[1] Effective arc length is defined as the distance between the midpoint of the line through the electrode tips and the midpoint of the arc tube at the U-bend.

[2] This temperature may be exceeded if allowed by the lamp manufacturer. Consult lamp manufacturer's literature.

[3] This temperature may be exceeded if allowed by the lamp manufacturer. Consult lamp manufacturer's literature.

HOW TO UNRAVEL STANDARDS

Teaching Infrastructure Studies

Martha Lampland and Susan Leigh Star

■ The research group we assembled at the University of California, Irvine, in fall 2001 was intended to draw together scholars working on apparently similar issues but doing so in isolation or in reference to communities not communicating with one another. The success of our efforts emboldened us to proselytize, to attract new scholars to the study of formalizing practices. In winter quarter, 2004, we taught a seminar on standards and quantification for our students in the Science Studies Program at University of California, San Diego. The course was designed to examine two questions: *How* have empirical and theoretical studies of infrastructure been conducted, and what specific issues arise in the *teaching* of this subject? We were met by a challenging group of students, whose intelligence, curiosity, and ingenuity made the process extremely rewarding. We learned by following them through the mazes of research; all examples mentioned in this appendix come from their work.

Infrastructure is designed to become invisible as it is stabilized. So, too, the processes of standardization are a means, produced to be preeminently interstitial and so fade into obscurity as they become more generally adopted. These idealized conditions are difficult to achieve, yet the political and technical struggles for invisibility central to infrastructural development and processes of standardization pose a significant challenge to their investigation. Hence, the study of infrastructure is quintessentially interdisciplinary, often global in scope, and very long-term in perspective. At the same time, the methods of history, legal studies, sociology, anthropology, and information science are not always easily meshed in attempting to recover the complex social and cultural techniques that make standardization and infrastructure possible. Although extremely productive as a research program, this is a difficult research program to teach in a ten-week quarter. We devised an approach that aimed at sidestepping the particular theoretical and methodological problems raised by infrastructure studies. This consists of foregrounding the object of investigation, that is, a tool or artifact of standardization (e.g., object, artifact, organism, material property, quality, or indicator). Following the trail of the object or disentangling the tendrils of the artifact forces the investigator to continually interrogate the imbrication of theory and evidence without being forced to choose the strictures of linearity, create simplifying models, or forgo the dynamics of history. In short, this required students to widen their perspective in the course of the quarter, resisting the kind of closure demanded in standard research narratives. Indeed, by moving backward and outward, students had a quick and easy lesson in what most of us actually do in the process of research: situate and define an object of analysis.

STUDENTS' PROJECTS

Sophia Efstathiou	Standards for dietary supplements
Mary Finley	Estimating fish population
Laura Harkewicz	Standards for radiation exposure
Steven Luis	Water-quality standards
Jesse Richmond	Categorization of African languages
Ray-Shyng Chou	GSM (Global System for Mobile communications) guidelines governing mobile technology
Michael Evans	Stellar magnitude scale
Lyn Headley	Textual citations in science
Brian Lindseth	JAVA
Eric Martin	*Caenorhabditis elegans*
Evelyn Range	Most recent farm bill passed by the U.S. Congress
Thomas Waidzunas	Electricity grid
Kevin Walsh	Science standards in California high schools
Mehmet Yalcinkaya	DSM (*Diagnostic and Statistical Manual of Mental Disorders*)

Communications 225B/History 239/Philosophy 209B/Sociology 255B (winter quarter, 2004)

Standardization and Quantification (Star/Lampland)

This course will consider standardization and quantification as a central issue in science studies. Scientific and cultural built environments are permeated with quantification and standards. Building codes, diagnostic categories, wage scales, clothing sizes, food safety thresholds, algorithms for assessing life expectancy, educational testing, and e-mail protocols are a handful of examples among millions that structure modern life. The group will examine the material aspects of standards and formalisms as embedded in infrastructure, and the historically specific circumstances of their birth. This will include social justice and cultural narratives inscribed in these technologies and arrangements. An immediate, central concern is how quantification, formalization, and standardization are culturally achieved. This includes questions about how techniques of quantification are discovered and disputed; the negotiation of particular standards and categories; and the experience of encountering standards, categories, and formalisms in the built and information environments (either as one disenfranchised or empowered by them). An open question and mandate for the seminar is to ask very broad questions about the imbrication of standards and metrics in everyday modern life. For example, how are aesthetics used to justify forms of standardization? How do changes in beliefs about quantification occur? What ethical principles are invoked (or ignored) in the standardizing process? We will consider both indigenous, local formalisms (both Western industrial and traditional, and their mixtures) and globalization. Questions include ethical implications of quantification and formalization, social histories of metrication, social forces of change in regimes of quantification and standardization, and associated work practices.

Requirements

We envision this course as a collective effort, dependent on the active participation of all members in one another's work: in face-to-face discussions, more private musings, and online debates. Grades will be based on class participation (broadly defined) and on a written class project.

Weekly Readings

Students will be expected to prepare to discuss weekly readings by bringing questions, issues, and problems they have to class, although we will be responsible for initiating class discussion of texts. Each week, one person will offer to be the class scribe, taking notes on the discussion and posting them afterward on the Web board. In the interest of keeping the conversation going beyond the classroom, we will encourage everyone to review class notes on the Web board, amend postings, develop ideas further, or bring up new ideas for discussion. (If there are not enough class times for all students to take turns being a scribe, other tasks may also be assigned, e.g., compiling a final bibliography from our extended readings.)

Class Project

Each student will be expected to develop a project to investigate a *tool* of standardization or quantification. By *tool* we mean an object, artifact, organism, material property, quality, or indicator that is deployed in scientific analysis. In the course of the quarter, students will be expected to submit four progress reports (at least five pages each) describing in greater and greater detail the features of that tool as revealed in their explorations. Each progress report should include:

- Analytic findings at each stage of investigation
- Ethics and values at issue
- Tie-ins with readings and class discussion
- Comments on the process of discovery
- A list of references consulted
- An annotated bibliography

Students should decide on their "tool of choice" by the second meeting of class (January 13). Progress reports shall be posted on the Web board by 5 p.m. on the following four Fridays: January 23, February 6, February 20, and March 5.

At the consecutive class meeting, one-half of our time will be devoted to presentations of students' projects. Our final class session will be fully devoted to a final ten-minute presentation of each student's project.

Course Schedule

January 6 General introduction, discussion of interests and requirements

January 13 Introductory Thoughts
  Bowker, "Information Mythology"
  Busch, "The Moral Economy of Grades and Standards"
  Pargman and Palme, "ASCII Imperialism"
  Zeruvabel, "Temporal Regularity" and "Schedules"

January 20 Infrastructures and Bureaucracies
  Barry, "Harmonised States"
  Latour, "Centers of Calculation"
  Thevenot, "Rules and Implements: Investment in Forms"
  Yates, "Communication Technology and the Growth of Internal Communication"

January 27 Measuring People
  Briggs, "Local Numbers and Global Power: The Role of Statistics"
  Cowan, excerpts on clothing size and standards from *More Work for Mother*
  Moore and Clarke, "Genital Conventions and Trangressions"
  Stern, "From Mestizophilia to Biotypology: Racialization and Science in Mexico, 1920–1960"
  Examine website for the Intersex Society of North America (http://isna.org/index.html)

February 3 Treating People
Casper and Clarke, "Making the Pap Smear into the Right Tool for the Job"
Epstein, "Beyond the Standard Human?"
Timmermans and Berg, "Introduction" and "Standardizing Risk" (Chapter 6)
Wiener, "Formalizing the Accountability Endeavor"

February 10 The Moral Order of Accounting
Lakoff, "Adaptive Will: The Evolution of Attention Deficit Disorder"
Lengwiler, "Double Standards: The History of Standardizing Humans in Modern Life Insurance"
Miller, "Accounting and Objectivity: The Invention of Calculating Selves and Calculable Spaces"
Rose, "Calculable Minds and Manageable Individuals"

February 17 Navigating Standards
Dunn, "Trojan Pig"
Collier, "The Intransigence of Things: Market Adjustment and the Problem of Pipes"
Harvey and Chrisman, "Boundary Objects and the Social Construction of GIS Technology"
Kirk and Kutchins, *The Selling of the DSM*
Saunders, "Standards in Education"

February 24 Becoming Numbers
Daston, "Enlightenment Calculations"
Desrosières, "How Real Are Statistics?"
Mirowski, "Looking for Those Natural Numbers: Dimensionless Constants and the Idea of Natural Measurement"
Pentland, "Getting Comfortable with the Numbers: Micro Production of Macro Order"

March 2 Formalisms
Beck, "What Doesn't Fit: The 'Residual Category' as Analytic Resource"
Cronon, "Pricing the Future: Grain"
Goguen, "Towards a Social, Ethical Theory of Information"

March 9 Final Reports

Bibliography

Barry, Andrew. 2001. Chapter 3 in *Political Machines: Governing a Technological Society*. London: Athlone Press.

Beck, Eevi E. 2003. "What Doesn't Fit: The 'Residual Category' as Analytic Resource." In *Social Thinking-Software Practice,* edited by Yvonne Dittrich, Christiane Floyd, and Ralf Klischewski, 161–79. Cambridge, Mass.: MIT Press.

Bowker, Geoffrey C. 1994. "Information Mythology: The World of/as Information." In *Information Acumen: The Understanding And Use Of Knowledge In Modern Business,* edited by Lisa Bud-Frierman, 231–47. London: Routledge.

Briggs, Charles, with Clara Mantini-Briggs. 2003. *Stories in the Time of Cholera: Racial Profiling during a Medical Nightmare.* Berkeley: University of California Press. Chapter 11.

Busch, Lawrence. 2000. "The Moral Economy of Grades and Standards." *Journal of Rural Studies* 16:273–83.

Casper, Monica, and Adele Clarke. 1998. "Making the Pap Smear into the Right Tool for the Job: Cervical Cancer Screening in the U.S., 1940–1995." *Social Studies of Science* 28:255–90.

Collier, Stephen. 2001. "The Intransigence of Things: Market Adjustment and the Problem of Pipes." In "Post-Socialist City: The Government of Society in Neo-Liberal Times." PhD diss., University of California, Berkeley.

Cowan, Ruth Schwartz. 1983. *More Work for Mother: The Ironies of Household Technology from the Open Hearth to the Microwave.* New York: Basic Books.

Cronon, William. 1991. Chapter 3 in Nature's Metropolis: Chicago and the Great West. New York: W. W. Norton.

Daston, Lorraine. 1994. "Enlightenment Calculations." *Critical Inquiry* 21:182–202.

Desrosières, A. 2001. "How Real Are Statistics?: Four Possible Attitudes." *Social Research.* 68(2): 339–55.

Dunn, Elizabeth. 2003. "Trojan Pig: Paradoxes of Food Safety Regulation." *Environment and Planning A* 35(8): 1493–1511.

Epstein, Steve. n.d. "Beyond the Standard Human?" Manuscript.

Goguen, Joseph. 1997. "Towards a Social, Ethical Theory of Information." In *Social Science Research, Technical Systems and Cooperative Work,* edited by Geoffrey Bowker, Les Gasser, Leigh Star, and William Turner, 27–56. Hillsdale, N.J.: Erlbaum. PDF version available at: http://www-cse.ucsd.edu/users/goguen/ps/sti.pdf.

Harvey, Francis, and Nicholas Chrisman. 1998. "Boundary Objects and the Social Construction of GIS Technology." *Environment and Planning A* 30(9): 1683–94.

Kirk, Stuart A., and Herb Kutchins. 1992. *The Selling of the DSM: The Rhetoric of Science in Psychiatry.* New York: Aldine de Gruyter.

Lakoff, Andrew. 2000. "Adaptive Will: The Evolution of Attention Deficit Disorder." *Journal of the History of the Behavioral Sciences* 36(2): 149–69.

Latour, Bruno. 1987. Chapter 6 in *Science in Action.* Cambridge, Mass.: Harvard University Press.

Lengwiler, Martin. n.d. "Double Standards: The History of Standardizing Humans in Modern Life Insurance." Manuscript.

Miller, Peter. 1992. "Accounting and Objectivity: The Invention of Calculating Selves and Calculable Spaces." *Annals of Scholarship* 9:61–86.

Mirowski, Philip. 1992. "Looking for Those Natural Numbers: Dimensionless Constants and the Idea of Natural Measurement." *Science in Context* 5(1): 165–88.

Moore, Lisa Jean, and Adele Clarke. 1995. "Genital Conventions and Trangressions: Graphic Representations in Anatomy Texts, c1900–1991." *Feminist Studies* 22(1): 255–301.

Pargman, Daniel, and Jacob Palme. n.d. "ASCII Imperialism." Manuscript.

Pentland, Brian T. 1993. "Getting Comfortable with the Numbers: Micro Production of Macro Order." *Accounting, Organizations and Society* 18(7/8): 605–20.

Rose, Nikolas. 1988. "Calculable Minds and Manageable Individuals." *History of the Human Sciences* 1:179–200.

Saunders, L.K. (Mimi). 2001. "Standards in Education." Manuscript.

Stern, Alexandra Minna. 2003. "From Mestizophilia to Biotypology: Racialization and Science in Mexico, 1920–1960." In *Race and Nation in Modern Latin America,* edited by Nancy P. Applebaum, Anne S. Macpherson, and Karin Alejandra Rosemblatt, 187–210. Chapel Hill: University of North Carolina Press.

Thévenot, Laurent. 1984. "Rules and Implements: Investment in Forms." *Social Science Information* 23(1): 1–45.

Timmermans, Stefan, and Marc Berg. 2003. *The Gold Standard: The Challenge of Evidence-Based Medicine and Standardization in Health Care.* Philadelphia: Temple University Press. Introduction and Chapter 6.

Wiener, Carolyn. 2000. *The Elusive Quest: Accountability in Hospitals.* New York: Aldine de Gruyter. Chapter 3.

Yates, JoAnne. 1989. Chapter 3 in *Control through Communication: The Rise of System in American Management.* Baltimore: John Hopkins U. Press.

Zeruvabel, Eviatar. 1981. *Hidden Rhythms: Schedules and Calendars in Social Life.* Chicago: University of Chicago Press. Chapters 1 and 2.

REFERENCES

ARCHIVAL MATERIAL

Archives Zurich Financial Services, Switzerland.
Declassified Secret Papers Collection (FMTÜK)
Győr-Moson-Sopron County Archive (GY-S-M)
Hajdú-Bihar County Archive (HBML)
Hungarian Workers' Party (Magyar Dolgozók Pártja, MDP): 1948–1956
Ministry of Agriculture: 1948–1956
Minutes of the Political Committee (MDP-PB)
Zala County Archive (ZML)

PUBLISHED MATERIAL

Aabakken, Olav. 1927. "On the Insurance of Under-Average Lives in Norway." In *Transactions of the 8th International Congress of Actuaries, London 27th to 30th June, 1927*, vol. 2, compiled by S.H. Jarvis, 143–50. London: Layton.
Abbate, Janet. 1999. *Inventing the Internet*. Cambridge, Mass.: MIT Press.
Aereboe, Friedrich. 1930. *Wirtschaft und Kultur in den Vereinigten Staaten von Nordamerika*. Berlin: Verlagsbuchhandlung Paul Parey.
Akrich, Madeleine. 1993. "Inscription et coordination socio-techniques: Anthropologie quelques dispositifs énergétiques." PhD diss., University of Lille III.
Altman, Lawrence K. 1987. *Who Goes First?: The Story of Self-Experimentation in Medicine*. New York: Random House.
Arensberg, Conrad M., and Solon T. Kimball. 1968. *Family and Community in Ireland*. Cambridge, Mass.: Harvard University Press.
Ariès, Philippe. 1962. *Centuries of Childhood: A Social History of Family Life*. New York: Alfred A. Knopf.
Arps, Ludwig. 1965. *Auf sicheren Pfeilern: Deutsche Versicherungswirtschaft vor 1914*. Göttingen: Vandenhoeck and Ruprecht.

———. 1970. *Durch unruhige Zeiten: Deutsche Versicherungswirtschaft seit 1914*, vol. 1, *Erster Weltkrieg und Inflation*. Karlsruhe: Verlag Versicherungswirtschaft.

Arzberger, Peter W., Peter Schroeder, Anne Beaulieu, Geoffrey C. Bowker, Kathleen Casey, Leif Laaksonen, David Moorman, Paul Uhlir, and Paul Wouters. 2004. "Promoting Access to Public Research Data for Scientific, Economic, and Social Development." *Data Science Journal* 3: 135–52.

Auerbach, Judith D., and Anne E. Figert. 1995. "Women's Health Research: Public Policy and Sociology." *Journal of Health and Social Behavior* 35 (supplement): 115–31.

Baird, Karen L. 1999. "The New NIH and FDA Medical Research Policies: Targeting Gender, Promoting Justice." *Journal of Health Politics, Policy and Law* 24(3): 531–65.

Baker, Karen S., and Helena Karasti. 2005. *The Long Term Information Management Trajectory: Working to Support Data, Science and Technology*. SIO Report, University of California, San Diego.

Baker, Karen S., and Florence Millerand. 2007. "Articulation Work Supporting Information Infrastructure Design: Coordination, Categorization, and Assessment in Practice." *Proceedings of the Hawaii International Conference on System Sciences* (CD-ROM), edited by Ralph H. Sprague (10 pages). Hawaii: Computer Society Press.

Bárczi, Géza. 1941. Magyar szófejtő szótár [Hungarian Etymological Dictionary]. Budapest: Királyi Magyar Egyetemi Nyomda.

Barry, Andrew. 2001. *Political Machines: Governing a Technological Society*. London: Athlone Press.

Baruch, Bernard. 1955. *Simpson's Contemporary Quotations*, compiled by James B. Simpson, 1988. Bartleby.com. http:www.bartleby.com/63/77/4477.html. Also available in *Simpson's Contemporary Quotations: The Most Notable Quotations, 1950–1988*, compiled by James B. Simpson. Boston: Houghton Mifflin, 1988.

Becker, Howard S. 1982. *Art Worlds*. Berkeley: University of California Press.

Bereznai, Aurél. 1943. *Munkaerőtervgazdálkodás a mezőgazdaságban* [Labor Resources in Agriculture in the Planned Economy]. Magyar Közigazgatástudományi Intézet 4. szám.

Berg, Marc. 1997. *Rationalizing Medical Work: Decision-Support Techniques and Medical Practices*. Cambridge, Mass.: MIT Press.

Bingen, Jim and Lawrence Busch. 2006. *Agricultural Standards: The Shape of the Global Food and Fiber System*. New York: Springer.

Bird, Chloe E. 1994. "Women's Representation as Subjects in Clinical Studies: A Pilot Study of Research Published in JAMA in 1990 and 1992." In *Women and Health Research: Ethical and Legal Issues of Including Women in Clinical Studies*, edited by Anna C. Mastroianni, Ruth Faden, and Daniel Federman, 151–73. Washington, D.C.: National Academy Press.

Blakely, Robert L., and Judith M. Harrington. 1997. "Grave Consequences: The Opportunistic Procurement of Cadavers at the Medical College of Georgia." In *Bones in the Basement: Postmortem Racism in 19th Century Medical Training*, edited by Robert L. Blakely and Judith M. Harrington, 162–83. Washington, D.C.: Smithsonian Institution Press.

Blantz, Jenő. 1936. "Megtakarítások a munkaköltségeknél a belterjesség szempontjából" [Savings in Labor Costs from the Point of View of Intensification]. In *A belterjes gazdálkodás lényege, jelentősége és előfeltételei*, edited by the Országos Magyar Gazdasági Egyesület, 99–121. Budapest: "Pátria" Irodalmi Vállalat és Nyomdai Részvénytársaság.

Blaschke, Ernst. 1895. *Denkschrift zur Lösung des Problems der Versicherung minderwertiger Leben*. Wien: Verlag des Lebensversicherungs-Teilungsvereines.

Bould, Sally, Beverly Sanborn, and Laura Reif. 1989. *Eighty-Five Plus: The Oldest Old*. Belmont, Calif.: Wadsworth Publishing.

Bourdieu, Pierre. 1984. *Distinction: A Social Critique of the Judgement of Taste*. Translated by Richard Nice. Cambridge, Mass.: Harvard University Press.

———. 1986. "The Forms of Capital." In *Handbook of Theory and Research for the Sociology of Education*, edited by John Richardson, 241–58. New York: Greenwood Press.

Bowker, Geoffrey. 1994a. "Information Mythology and Infrastructure." In *Information Acumen: The Understanding and Use of Knowledge in Modern Business*, edited by Lisa Bud-Frierman, 231–47. London: Routledge.

———. 1994b. *Science on the Run: Information Management and Industrial Geophysics at Schlumberger, 1920–1940*. Cambridge, Mass.: MIT Press.

———. 2006. *Memory Practices in the Sciences*. Cambridge, Mass.: MIT Press.

Bowker, Geoffrey C., and Susan Leigh Star. 1999. *Sorting Things Out: Classification and Its Consequences*. Cambridge, Mass.: MIT Press.

Braverman, Harry. 1974. *Labor and Monopoly Capital: The Degradation of Work in the Twentieth Century*. New York: Monthly Review Press.

Briggs, Charles, with Clara Mantini-Briggs. 2003. *Stories in the Time of Cholera: Racial Profiling during a Medical Nightmare*. Berkeley: University of California Press.

Broberg, Gunnar, ed. 1996. *Eugenics and the Welfare State: Sterilization Policy in Denmark, Sweden, Norway, and Finland*. (Uppsala Studies in History of Science 21.) East Lansing: Michigan State University.

Brown, Chester T. 1935. "Methods of Estimating Risks." In *International Congress of Life Assurance Medicine, 94–99*. London.

Brown, Nik, and Mike Michael. 2001. "Switching between Science and Culture in Transpecies Transplantation." *Science, Technology, and Human Values* 26(1): 3–22.

Brown, Warren. 1999. "Developing a Crash Course on Safety: With Test Dummies, Engineering Firm Teams Up with Honda to Reduce Injuries to Pedestrians." *Washington Business (Washington Post)*, 12 April, 10–11.

Brunsson, Nils, and Bengt Jacobsson and Associates. 2000. *A World of Standards*. Oxford: Oxford University Press.

Buchheim, Eduard. 1878. *Handbuch für Versicherungs-Aerzte: Aerztliche Versicherungskunde*. Leipzig: Vogel.

———. 1887. *Ärztliche Versicherungs-Diagnostik*. Wien: Hölder.

Budd, John W., and Timothy Guinnane. 1991. "Intentional Age-Misreporting, Age-Heaping, and the 1908 Old Age Pensions Act in Ireland." *Population Studies* 45: 497–518.

Busch, Lawrence. 2000. "The Moral Economy of Grades and Standards." *Journal of Rural Studies* 16: 273–83.

Busch, Lawrence, Keiko Tanaka, and Valerie J. Gunter. 2000. "Who Cares If the Rat Dies?: Rodents, Risks, and Humans in the Science of Food Safety." In *Illness and the Environment: A Reader in Contested Medicine*, edited by Steve Kroll-Smith, Phil Brown, and Valerie J. Gunter, 108–19. New York: New York University Press.

Callon, Michel, ed. 1998. *The Laws of the Markets*. Malden, Mass.: Blackwell Publishers/The Sociological Review.

Carlin, George. 2008. Accessed at "George Carlin on Age." *Thoughts . . . Postcards from the Heart . . .* http://thoughts.sdwhiteonline.com/2008/03/23/george-carlin-on-age/.

Cartwright, Lisa. 1997. "The Visible Man: The Male Criminal Subject as Biomedical Norm." In *Processed Lives: Gender and Technology in Everyday Life*, edited by Jennifer Terry and Melodie Calvert, 123–37. London: Routledge.

Chrisman, Nicholas. 1997. *Exploring Geographic Information Systems*. New York: J. Wiley and Sons.

Chudacoff, Howard P. 1989. *How Old Are You?: Age Consciousness in American Culture*. Princeton: Princeton University Press.

Clark, Geoffrey. 1999. *Betting on Lives: The Culture of Life Insurance in England 1695–1775*. Manchester, UK: Manchester University Press.

Clarke, Adele E. 1995. *Human Materials as Contested Objects: Problematics of Subjects Who Speak*. Unpublished manuscript, University of California, San Francisco.

———. 1998. *Disciplining Reproduction: Modernity, American Life Sciences and the "Problem of Sex."* Berkeley: University of California Press.

Clarke, Adele E., and Joan H. Fujimura, eds. 1992a. *The Right Tools for the Job: At Work in Twentieth-Century Life Sciences*. Princeton: Princeton University Press.

———. 1992b. "What Tools? Which Jobs? Why Right?" In *The Right Tools for the Job: At Work in Twentieth-Century Life Sciences*, edited by Adele E. Clarke and Joan H. Fujimura, 3–43. Princeton: Princeton University Press.

CNN. 2008. "Euromyths: Facts and Fiction in EU law." CNN online edition, 18 January. Available at: edition.cnn.com/specials/2000/eurounion.

Cohen, Jay S. 2001. *Overdose: The Case against the Drug Companies*. New York: Penguin Putnam.

Cole, Thomas R. 1992. *The Journey of Life: A Cultural History of Aging in America*. Cambridge, UK: Cambridge University Press.

Collier, Stephen. 2001. "The Intransigence of Things: Market Adjustment and the Problem of Pipes." In "Post-Socialist City: The Government of Society in Neo-Liberal Times." PhD diss., University of California, Berkeley.

Collins, Patricia Hill. 1990. *Black Feminist Thought: Knowledge, Consciousness and the Politics of Empowerment*. New York: Routledge.

Conrad, Peter. 2005. "The Shifting Engines of Medicalization." *Journal of Health and Social Behavior* 46(1): 3–14.

Conrad, Peter, and Joseph W. Schneider. 1980. *Deviance and Medicalization: From Badness to Sickness*. St. Louis: C. V. Mosby.

Coppet, L. 1935. Contribution to discussion. In *International Congress of Life Assurance Medicine*, edited by Otto May, 106–109. London.

Corbie-Smith, Giselle. 1999. "The Continuing Legacy of the Tuskegee Syphilis Study: Considerations for Clinical Investigation." *American Journal of the Medical Sciences* 317(1): 5–8.

Corrigan, Oonagh P. 2002. "A Risky Business: The Detection of Adverse Drug Reactions in Clinical Trials and Post-Marketing Exercises." *Social Science and Medicine* 55: 497–507.

Costa, Dora L. 1998. *The Evolution of Retirement: An American Economic History, 1880–1990*. Chicago: University of Chicago Press.

Crane, Diane. 1972. *Invisible Colleges: Diffusion of Knowledge in Scientific Communities*. Chicago: University of Chicago Press.

Cravens, Hamilton. 1978. *The Triumph of Evolution: American Scientists and the Heredity-Environment Controversy, 1900–1941*. Philadelphia: University of Pennsylvania Press.

Cronon, William. 1991. *Nature's Metropolis: Chicago and the Great West*. New York: W. W. Norton.

Crystal, David. 1997. *English as a Global Language*. Cambridge, UK: Cambridge University Press.

Darnton, John. 1994. "The Bent-Banana Ban and Other British Gibes at Europe." *New York Times,* October 6, 1994, 16.

Daston, Lorraine J. 1988. *Classical Probability in the Enlightenment*. Princeton: Princeton University Press.

Daston, Lorraine J., and Peter Galison. 1992. "The Image of Objectivity." *Representations* 40: 81–128.

David, Paul. 1985. "Clio and the Economics of QWERTY." *American Economic Review* 75: 332–37.

Davis, Audrey B. 1981. "Life Insurance and the Physical Examination: A Chapter in the Rise of American Medical Technology." *Bulletin of the History of Medicine* 55: 392–406.

Degler, Carl N. 1991. *In Search of Human Nature: The Decline and Revival of Darwinism in American Social Thought*. New York: Oxford University Press.

Desrosières, Alain. 2001. "How Real are Statistics? Four Possible Attitudes." *Social Research* 68: 339–55.

Dominguez, Virginia. 1994. "A Taste for 'the Other': Intellectual Complicity in Racializing Practices." *Current Anthropology* 35(4): 333–348.

Douglas, Mary. 1986. *How Institutions Think*. Syracuse: Syracuse University Press.

Dowbiggin, Ian R. 1985. "Degeneration and Hereditarianism in French Mental Medicine, 1840–1890." In *The Anatomy of Madness: Essays in the History of Psychiatry*, vol. 1, edited by William F. Bynum, Roy Porter, and Michael Shepherd, 188–233. London: Tavistock.

———. 1991. *Inheriting Madness: Professionalization and Psychiatric Knowledge in Nineteenth-Century France*. (Medicine and Society 4.) Berkeley: University of California Press.

———. 1997. *Keeping America Sane: Psychiatry and Eugenics in the United States and Canada, 1880–1940*. (Cornell Studies in the History of Psychiatry.) Ithaca: Cornell University Press.

Drachmann, Bj. 1927. "Insurance of Under-Average Lives—Methods and Experiences in Denmark." In *Transactions of the 8th International Congress of Actuaries, London 27th to 30th June, 1927*, vol. 2, compiled by S. H. Jarvis, 120–27. London: Layton.

Dresser, Rebecca. 1992. "Wanted: Single White Male for Medical Research." *Hastings Center Report* (January–February): 24–29.

Duenwald, Mary. 2002. "Hormone Therapy: One Size, Clearly, No Longer Fits All." *New York Times*, 16 July, D-1, D-6.

———. 2003. "One Size Definitely Does Not Fit All." *New York Times*, 22 June, sec.15, Page 1.

Duster, Troy. 1990. *Backdoor to Eugenics*. New York: Routledge.

Edwards, Paul N. 1996. *The Closed World: Computers and the Politics of Discourse in Cold War America*. Cambridge, Mass.: MIT Press.

Elster, Ludwig. 1880. *Die Lebensversicherung in Deutschland: Ihre volkswirtschaftliche Bedeutung und die Notwendigkeit ihrer gesetzlichen Regelung*. Halle: Frommann.

Epstein, Steven. 1996. *Impure Science: AIDS, Activism, and the Politics of Knowledge*. Berkeley: University of California Press.

———. 1997. "Activism, Drug Regulation, and the Politics of Therapeutic Evaluation in the AIDS Era: A Case Study of ddC and the 'Surrogate Markers' Debate." *Social Studies of Science* 27(5): 691–726.

———. 2003. "Sexualizing Governance and Medicalizing Identities: The Emergence of 'State-Centered' LGBT Health Politics in the United States." *Sexualities* 6(2): 131–71.

——. 2004. "Bodily Differences and Collective Identities: The Politics of Gender and Race in Biomedical Research in the United States." *Body and Society* 10(2–3): 183–203.

——. 2005. "Institutionalizing the New Politics of Difference in U.S. Biomedical Research: Thinking across the Science/State/Society Divides." In *The New Political Sociology of Science: Institutions, Networks and Power*, edited by Scott Frickel and Kelly Moore, 327–50. Madison: University of Wisconsin Press.

——. 2007. *Inclusion: The Politics of Difference in Medical Research*. Chicago: University of Chicago Press.

Erdei, Ferenc. 1941. *A magyar paraszttársadalom* [Hungarian Peasant Society]. Budapest: Franklin Társulat.

European Commission. 2008a. "Euromyths: Curved Bananas." European Commission's Press Office in London, 18 January. Available at: http://ec.europa.eu/unitedkingdom/press/euromyths/myth05_en.htm.

——. 2008b. "Euromyths—Keyword Index." European Commission's Press Office in London, 18 January. Available at: http://ec.europa.eu/unitedkingdom/press/euromyths/index_en.htm.

Ewald, François. 1986. *L'Etat Providence*. Paris: Editions Grasset and Fasquelle.

Fadiman, Anne. 1997. *The Spirit Catches You and You Fall Down*. New York: Farrar, Straus and Geroux.

Fischer, Andor. 1936. "Munkatanulmányok és költségszámitások a burgonyatermesztés köréből" [Work Studies and Cost Accounting from the Domain of Potato Cultivation]. PhD diss., M. Kir. József Nádor Műszaki és Gazdaságtudományi Egyetem.

Fischer, David Hackett. 1978. *Growing Old in America*. New York: Oxford University Press.

Fleming, James. 1998. *Historical Perspectives on Climate Change*. New York: Oxford University Press.

Florschütz, Georg. 1925. "Lebensversicherungsmedizin 1900–1924." *Zeitschrift für die gesamte Versicherungswissenschaft* 25: 46–52.

Fodor, Zsigmond. 1925. "A Taylor-rendszer és a mezőgazdaság" [The Taylor System and Agriculture]. *Mezőgazdasági Üzemtani Közlemények* 3(3–4): 42–49.

Food and Drug Administration. "FDA Approves Humatrope for Short Stature 2003." FDA talk paper. Available at: http://www.fda.gov/opacom/hpwhats.html. Accessed August 1, 2003.

Foster, Morris W. 2003. "Pharmacogenomics and the Social Construction of Identity." In *Pharmacogenomics: Social, Ethical, and Clinical Dimensions*, edited by Mark A. Rothstein, 251–65. Hoboken: John Wiley and Sons.

Foucault, Michel. 1979. *Discipline and Punish: The Birth of the Prison*. New York: Vintage Books.

——. 1991. "Governmentality." In *The Foucault Effect: Studies in Governmentality*, edited by Graham Burchell, Colin Gordon, and Peter Miller, 87–104. Chicago: University of Chicago Press.

Fox, Margalit. 2005. "Samuel Alderson, Crash-Test Dummy Inventor, Dies at 90." *New York Times*, 18 February, A21.

Franklin, Jerry F., Caroline S. Bledsoe, and James T. Callahan. 1990. "Contributions of the Long-Term Ecological Research Program." *BioScience* 40(7): 509–15.

Friedland, William H., Amy E. Barton, and Robert J. Thomas. 1981. *Manufacturing Green Gold: Capital, Labor, and Technology in the Lettuce Industry*. Cambridge, UK: Cambridge University Press.

Friedman, Batya. 1996. "Value-Sensitive Design." *Interactions of the ACM* 30: 17–23.

Friedman, Batya, and Nissenbaum, Helen. 1997. "Bias in Computer Systems." In *Human Values and the Design of Computer Technology*, edited by Batya Friedman, 21–40. New York: Cambridge University Press.

Fujimura, Joan H. 1991. "On Methods, Ontologies, and Representation in the Sociology of Science: Where Do We Stand?" In *Social Organization and Social Process*, edited by David. R. Maines, 207–48. New York: Aldine de Gruyter.

——. 1992. "Crafting Science: Standardized Packages, Boundary Objects, and Translation." In *Science as Practice and Culture*, edited by Andrew Pickering, 168–214. Chicago: University of Chicago Press.

Gárdonyi, Sándor. 1933. Az energiaeloszlás elméletének alaptételei és grafikai ábrázolása vállalatok szervezésénél [Basic Principles of the Theory of the Dispersion of Energy and its Graphic Representation in the Organization of Enterprises]. *Mezőgazdasági Üzem és Számtartás* 5(2):1–3, 32–36.

Garfinkel, Perry. 2001. "Medical Students Get Taste of Real-Life Doctoring." *New York Times*, 23 October, D-7.

Gasser, Les. 1986. "The Integration of Computing and Routine Work." *ACM Transaction on Office Information Systems* 4: 205–25.

Gauthier, Jason G. 2002. *Measuring America: The Decennial Censuses from 1790 to 2000*. Washington, D.C.: U.S. Census Bureau.

Gebauer, Max. 1895. *Die sogenannte Lebensversicherung*. (Staatswissenschaftliche Studien, vol. 5/3.) Jena: Fischer.

Gigerenzer, Gerd, Zeno Swijtink, Theodore Porter, Lorraine Daston, John Beatty, and Lorenz Krüger. 1989. *The Empire of Chance: How Probability Changed Science and Everyday Life*. Cambridge, UK: Cambridge University Press.

Gille, Zsuzsa. 2007. *From the Cult of Waste to the Trash Heap of History: The Politics of Waste in Socialist and Postsocialist Hungary*. Bloomington: Indiana University Press.

Gillis, John R. 1974. *Youth and History*. New York: Academic Press.

Gilman, Sander L. 1985. *Difference and Pathology: Stereotypes of Sexuality, Race and Madness*. Ithaca: Cornell University Press.

Glasson, William H. 1918. *Federal Military Pensions in the United States*. New York: Oxford University Press.

Glendon, Mary Ann. 1989. *The Transformation of Family Law: State, Law, and Family in the United States and Europe*. Chicago: University of Chicago Press.

Gordon, Deborah R. 1988. "Clinical Science and Clinical Expertise: Changing Boundaries between Art and Science in Medicine." In *Biomedicine Examined*, edited by Margaret Lock and Deborah R. Gordon, 257–95. Dordrecht: Kluwer Academic Publishing.

Gorman, Christine. 1999. "Drugs by Design." *Time*, 11 January, 78.

Gould, Stephen Jay. 1996. *The Mismeasure of Man*. New York: Norton.

Gratton, Brian. 1986. *Urban Elders: Family, Work, and Welfare among Boston's Aged, 1890–1950*. Philadelphia: Temple University Press.

Greene, Charles Lyman. 1905. *Medical Examination for Life Insurance and Its Associated Clinical Methods with Chapters on the Insurance of Substandard Lives and Accident Insurance*. London: Rebman.

Griebler, Thomas H. 1992. *Bearbeitung von erhöhten Risiken in der Lebensversicherung*. unpublished M.A. thesis, Wirtschaftsuniversität, Vienna.

Griesemer, James R. 1990. "Modeling in the Museum: On the Role of Remnant Models in the Work of Joseph Grinnell." *Biology and Philosophy* 5: 3–36.

Grimm, Jacob, and Wilhelm Grimm. 1885. *Deutsches Wörterbuch, Bd. 6* Leipzig.

Grimmer-Solem, Erik. 2003. *The Rise of Historical Economics and Social Reform in Germany, 1864–1894.* Oxford: Clarendon Press.

Gruder, O. 1927. "Die Versicherung anormaler Leben in Oesterreich." In *Transactions of the 8th International Congress of Actuaries, London 27th to 30th June, 1927,* Vol. 4, compiled by S.H. Jarvis, 283–307. London: Layton.

Guillet de Monthoux, Pierre. 1981. *Doktor Kant och den oekonomiska rationaliseringen* [Doctor Kant and the Uneconomic Rationalization]. Göteborg: Korpen.

Györffy, István. 1942. *Magyar nép, magyar föld* [Hungarian People, Hungarian Land]. Budapest: Turul Kiadás.

Haber, Carole.1983. *Beyond Sixty-Five: The Dilemma of Old Age in America's Past.* Cambridge, UK: Cambridge University Press.

Hacking, Ian. 1990. *The Taming of Chance.* Cambridge, UK: Cambrigde University Press.

Haegler, Adolf. 1896. *Über die Factoren der Widerstandskraft und die Vorhersage der Lebensdauer beim gesunden Menschen.* Basel: Schwabe.

Hannemyr, Gisle. 1999. "Technology and Pleasure: Considering Hacking Constructive." *First Monday* 4(2). Available at: http://www.firstmonday.org/issues/issue4_2/gisle/index .html.

Hanseth, Ole, Eric Monteiro, and Morten Hatling. 1996. "Developing Information Infrastructure: The Tension between Standardization and Flexibility." *Science, Technology, and Human Values* 21(4): 407–26.

Hatvani (Hofferik), István. 1935. *A pszichotechnika szerepe és jelentősége a munka racionalizálásában* [The Role and Significance of Psychotechnique in the Rationalization of Work]. Győr: Baross-nyomda.

Hedgecoe, Adam. 2004. *The Politics of Personalized Medicine: Pharmacogenetics in the Clinic.* Cambridge, UK: Cambridge University Press.

Hobbie, John E., Stephen R. Carpenter, Nancy B. Grimm, James R. Gosz, and Timoty R. Seastedt. 2003. "The US Long Term Ecological Research Program." *BioScience* 53(1): 21–32.

Hörnig, P. 1935a. "Discussion Vote." In *International Congress of Life Assurance Medicine,* edited by Otto May, 112. London.

———. 1935b. "Eroeffnungsrede." In *International Congress of Life Assurance Medicine,* edited by Otto May, 8. London.

———. 1935c. "Ueberblick ueber die geschichtliche Entwicklung und den Ausbau der Versicherung erhoehter Risiken." In *International Congress of Life Assurance Medicine,* edited by Otto May, 577–81. London.

Hughes, Thomas P. 1983. *Networks of Power: Electrification in Western Society, 1880–1930.* Baltimore: Johns Hopkins University Press.

———. 1989. "The Evolution of Large Technological Systems." In *The Social Construction of Technological Systems,* edited by Wiebe E. Bijker, Thomas P. Hughes, and Trevor Pinch, 51–82. Cambridge, Mass.: MIT Press.

Ikenberrry, G. John, and Theda Skocpol. 1987. "Expanding Social Benefits: The Role of Social Security." *Political Science Quarterly* 102: 389–416.

Illich, Ivan. 1981. *Shadow Work.* Boston: Marion Boyars.

Illyés, Gyula. 1936. *Puszták népe* [People of the Puszta]. Budapest: Nyugat Kiadó és Irodalmi R.T.

Jensvold, Margaret F., Jean A. Hamilton, and Billie Mackey. 1994. "Including Women in Clinical Trials: How About the Women Scientists?" *Journal of the American Medical Women's Association* 49(4): 110–12.

Jewett, Tom, and Rob Kling. 1991. "The Dynamics of Computerization in a Social Science Research Team: A Case Study of Infrastructure, Strategies, and Skills." *Social Science Computer Review* 9: 246–75.

Johnson, Ericka. 2005. "The Ghost of Anatomies Past: Simulating the One-Sex Body in Modern Medical Training." *Feminist Theory* 6(2): 141–59.

Jones, Matthew B., Chad Berkley, Jivka Bojilova, and Mark Schildhauer. 2001. "Managing Scientific Metadata." *IEEE Internet Computing* 5(5): 59–68.

Jordan, Kathleen, and Michael Lynch. 1998. "The Dissemination, Standardization and Routinization of a Molecular Biological Technique." *Social Studies of Science* 28(5–6): 773–800.

Josling, Timothy E., and Timothy G. Taylor, eds. 2003. *Banana Wars: The Anatomy of a Trade Dispute.* Cambridge, Mass.: CABI Publishing.

Kahn, Jonathan. 2004. "How a Drug Becomes 'Ethnic': Law, Commerce, and the Production of Racial Categories in Medicine." *Yale Journal of Health Policy, Law and Ethics* 4(1): 1–46.

Karasti, Helena, and Karen S. Baker. 2004. "Infrastructuring for the Long-Term: Ecological Information Management." *Proceedings of the Hawaii International Conference on System Sciences* (CD-ROM), edited by Ralph H. Sprague (10 pages). Hawaii: Computer Society Press.

Karasti, Helena, Karen S. Baker, and Eija Halkola. 2006. "Enriching the Notion of Data Curation in E-Science: Data Managing and Information Infrastructuring in the Long Term Ecological Research (LTER) Network." *Computer Supported Cooperative Work* 15(4): 321–58.

Károly, Rezső. 1924. "A munkaszervezet javítása és a Taylor-rendszer a mezőgazdaságban" [The Amelioration of Work Organization and the Taylor System in Agriculture]. *Mezőgazdasági Üzemtani Közlemények* 2(3–4): 75–86.

———. 1925. "A mezőgazdasági könyvelés szerepe és szervei" [The Role and Administrative Organs of Agricultural Accounting]. *Mezőgazdasági Üzemtani Közlemények* 3(1-2): 1-11.

Karup, Johannes, Robert Gollmer, and Georg Florschütz. 1902. *Aus der Praxis der Gothaer Lebensversicherungsbank: Versicherungs-Statistisches und -Medizinisches.* Jena: Gustav Fischer.

Kehm, Max. 1897. *Über die Versicherung minderwertiger Leben.* (Staatswissenschaftliche Studien, vol. 6/6.) Jena: G. Fischer.

Kemptner, Ernő. 1939 "A mezőgazdasági munka célszerűsítése és a szociális helyzet javítása" [The Rationalization of Agriculture Work and the Improvement of Social Conditions]. *Köztelek Zsebnaptár* 45(2): 27–36.

Kerék, Mihály. 1939. *A magyar földkérdés* [The Land Question in Hungary]. Budapest: Mefhosz Könyvkiadó.

Keszler, Gyula. 1941. "A teljesítménnyel arányos munkabérfizetés" [Wage Payment Proportional to Output]. *Közgazdasági Szemle* 65(84): 869–93.

Kevles, Daniel. 1968. "Testing the Army's Intelligence: Psychologists and the Military in World War I." *Journal of American History* 55: 565–81.

King, Patricia A. 1992. "The Dangers of Difference." *Hastings Center Report* 22(6): 35–38.

Koch, Julius Ludwig August. 1891–1893. *Die psychopathischen Minderwertigkeiten*, 3 vols. Ravensburg: Maier.

Kodar, Jenő. 1944. "Üzemvezetés és psychológia" [Firm Management and Psychology]. Weiss Manfred Vállalatok kulturális-szociális osztályának szabad egyetemén tartott előadása, 10 January.

Kohler, Robert E. 1994. *Lords of the Fly: Drosophila Genetics and the Experimental Life*. Chicago: University of Chicago Press.

Kohli, Martin.1986. "The World We Forgot: A Historical Review of the Life Course." In *Later Life: The Social Psychology of Aging*, edited by Victor W. Marshall, 271–303. Beverly Hills: Sage Publications.

Kolata, Gina. 1999. "Using Gene Testing to Decide a Patient's Treatment." *New York Times*, 20 December, A1.

———. 2001. " 'Maximum' Heart Rate Theory Is Challenged." *New York Times*, 24 April, D-1, D-8.

Kölber, László. 1932. *Munkatanulmányok a dunántúli tengeritermelés köréből* [Work Studies from the Domain of Corn Production on the Dunántúl]. Unpublished PhD dissertation, M. Kir. József Nádor Múszaki és Gazdaságtudományi Egyetem.

Kovács, Imre. 1937. *Néma forradalom* [Silent Revolution]. Budapest: Cserépfalvi Könyvkiadó.

Krach, Constance A., and Victoria A. Velkoff. 1999. "Centenarians in the United States, 1990." In *Current Population Reports* (Special Studies; P23–199RV), 1–18. Washington, D.C.: Government Printing Office.

Krauss, Michael. 1992. "The World's Languages in Crisis." *Language* 68(1): 4–10.

Krieger, Nancy, and Elizabeth Fee. 1996. "Man-Made Medicine and Women's Health: The Biopolitics of Sex/Gender and Race/Ethnicity." In *Man-Made Medicine: Women's Health, Public Policy, and Reform*, edited by Kasy L. Moss, 15–35. Durham: Duke University Press.

Kula, Witold. 1986. *Measures and Men*. Princeton: Princeton University Press.

Kulin, Sándor. 1933. "A modern földközösség" [The Modern Community of Land]. *Mezőgazdasági Üzem és Számtartás* 5(3–4): 56–61.

Lampland, Martha. 1995. *The Object of Labor: Commodification in Socialist Hungary*. Chicago: University of Chicago Press.

———. n.d.a. "The Allure of Models: Designing Wages to Modernize Hungarian Agriculture, 1920–1956." Unpublished manuscript, San Diego.

———. n.d.b. "Planning Economies: Science, Expertise and the Transition to Socialism in Hungary (1920–1956)." Under review for inclusion in *Technopolitics of Global Cold War*, edited by Gabrielle Hecht. Cambridge: MIT Press.

———. n.d.c. "Transforming Objects: The Transition from Feudalism to Capitalism in 19th Century Hungary." Unpublished manuscript, San Diego.

Lancaster, Roger N. 2003. *The Trouble with Nature: Sex in Science and Popular Culture*. Berkeley: University of California.

Latour, Bruno. 1987. *Science in Action: How to Follow Scientists and Engineers through Society*. Cambridge, Mass.: Harvard University Press.

———. 1993. *Aramis ou l'amour des techniques*. Paris: La Découverte.

———. 1996. *Aramis, or the Love of Technology*. Cambridge, Mass.: Harvard University Press.

———. 2005. *Reassembling the Social: An Introduction to Actor-Network-Theory*. Oxford: Clarendon.

Latour, Bruno, and Emilie Hernant. 1999. *Paris: Ville Invisible*. Paris: La Découverte.

Latour, Bruno, and Steve Woolgar. 1979. *Laboratory Life*. Thousands Oaks, Calif.: Sage Publications.

Lawrence, Susan C., and Kae Bendixen. 1992. "His and Hers: Male and Female Anatomy in Anatomy Texts for U.S. Medical Students, 1890–1989." *Social Science and Medicine* 35(7): 925–34.

Lederer, Ernst. 1914. *Entstehung, Entwickelung und heutiger Stand der Volksversicherung in Deutschland*. Erlangen.

Lehr, William. 1992. "Standardization: Understanding the Process." *Journal of the American Society for Information Science*. 43(8): 550–55.

Lengwiler, Martin. 2000. *Zwischen Klinik und Kaserne: Die Geschichte der Militärpsychiatrie in Deutschland und der Schweiz 1870–1914*. Zürich: Chronos.

———. 2003. "Psychiatry beyond the Asylum: The Origins of German Military Psychiatry before the First World War." *History of Psychiatry* 14(1): 41–62.

Lessig, Lawrence. 1999. *Code and Other Laws of Cyberspace*. New York: Basic Books.

Levien, Ralph. 1998. "The Decommoditization of Protocols." Available at: http://www .levien.com/free/decommoditizing.html.

Levy, Rene. 1996. "Toward a Theory of Life Course Institutionalization." In *Society and Biography*, edited by Ansgar Weyman and Walter R. Heinz, 83–108. Weinheim: Deutscher Studien Verlag.

Levy, Richard A. 1993. *Ethnic and Racial Differences in Responses to Medicines: Preserving Individualized Therapy in Managed Pharmaceutical Programs*. Reston, Va.: National Pharmaceutical Council.

Lewontin, Richard C. 2000. *The Triple Helix: Gene, Organism, and Environment*. Cambridge, Mass.: Harvard University Press.

Lexis, Wilhelm. 1906. "Die Volksversicherung." In *5. Internationaler Kongress für Versicherungs-Wissenschaft, Berlin, 10–15 September 1906*, vol. 1, edited by Alfred Manes, 21. Berlin: Mittler & Sohn.

Lin, Keh-Ming, Russell E. Poland, Michael W. Smith, Tony L. Strickland, and Ricardo Mendoza. 1991. "Pharmacokinetic and Other Related Factors Affecting Psychotropic Responses in Asians." *Psychopharmacology Bulletin* 27(4): 427–39.

Logan, John R., Russell Ward, and Glenna Spitze. 1992. "As Old as You Feel: Age Identity in Middle and Later Life." *Social Forces* 71: 451–67.

Löwy, Ilana, and Jean-Paul Gaudillière. 1998. "Disciplining Cancer: Mice and the Practice of Genetic Purity." In *The Invisible Industrialist: Manufactures and the Production of Scientific Knowledge*, edited by Jean-Paul Gaudillière and Ilana Löwy, 209–49. Houndsmills, UK: Macmillan.

Luke, Timothy W. 1999. "Environmentality as Green Governmentality." In *Discourses of the Environment*, edited by Eric Darier, 121–51. Oxford: Blackwell.

Lynch, Michael. 1991. "Pictures of Nothing: Visual Construals in Social Theory." *Sociological Theory* 9(1): 1–22.

MacKenzie, Donald. 2001. *Mechanizing Proof: Computing, Risk and Trust*. Cambridge, Mass.: MIT Press.

Maingie, Louis. 1927. "L'assurance des risques tares." In *Transactions of the 8th International Congress of Actuaries, London 27th to 30th June, 1927*, vol. 3, compiled by S. H. Jarvis, 262–276. London: Layton.

Manes, Alfred. 1913. *Versicherungswesen*. 2nd ed. (B.G. Teubners Handbücher für Handel und Gewerbe.) Leipzig, Berlin: Teubner.

Mark, James. 2005a. "Discrimination, Opportunity, and Middle-Class Success in Early Communist Hungary." *Historical Journal* 48(2): 499–521.

———. 2005b. "Society, Resistance and Revolution: The Budapest Middle Class and the Hungarian Communist State 1948–56." *English Historical Review* 488: 963–86.

Markel, Howard. 2002. "New Growth Chart Dispels the Myth That One Size Fits All." *New York Times*, 16 April, D5.

Marks, Harry M. 1997. *The Progress of Experiment: Science and Therapeutic Reform in the United States, 1900–1990.* Cambridge, UK: Cambridge University Press.

Mastroianni, Anna C., Ruth Faden, and Daniel Federman, eds. 1994. *Women and Health Research: Ethical and Legal Issues of Including Women in Clinical Studies.* vol. 1. Washington, D.C.: National Academy Press.

Matthews, J. Rosser. 1995. *Quantification and the Quest for Medical Certainty.* Princeton: Princeton University Press.

Mayr, Ernst and William Province, eds. 1980. *The Evolutionary Synthesis: Perspectives on the Unification of Biology.* Cambridge: Harvard University Press.

McCallum, Cecilia. 2005. "Racialized Bodies, Naturalized Classes: Moving through the City of Salvador da Bahia." *American Ethnologist* 32(1): 100–117.

Meinert, Curtis L. 1995. "The Inclusion of Women in Clinical Trials." *Science* 269(5225): 795–96.

Meinert, Curtis L., and Adele Kaplan Gilpin. 2001. "Estimation of Gender Bias in Clinical Trials." *Statistics in Medicine* 20: 1153–64.

Merton, Robert K. 1957. *Social Theory and Social Structure.* Glencoe: Free Press

Meyers Lexikon. 1939. vol. 7. Leipzig: Bibliographisches Institut.

Michener, William K., James W. Brunt, John J. Helly, Thomas B. Kirchner, and Susan G. Stafford. 1997. "Nongeospatial Metadata for the Ecological Sciences." *Ecological Applications* 7(1): 330–42.

Millerand, Florence. 1999. "Usages des NTIC, les approches de la diffusion, de l'innovation et de l'appropriation (2e partie)." *COMMposite* 99(1). Available at: http://commposite .org/v1/99.1/articles/ntic_2.htm.

Molnár, Ilona, and Pál Schiller. 1937. "Állandó ingerek hatása a munkára" [The Effect of Constant Stimuli on Work]. *Lélektani tanulmányok* 1: 22–29.

Morgan, Mary S., and Margaret Morrison, eds. 1999. *Models as Mediators: Perspectives on Natural and Social Science.* Cambridge, UK: Cambridge University Press.

Morris-Suzuki, Tessa. 2000. "Ethnic Engineering: Scientific Racism and Public Opinion Surveys in Midcentury Japan." *Positions* 8(2): 499–529.

Moser, Marvin. 1995. "Black-White Differences in Response to Antihypertensive Medication." *Journal of the National Medical Association* 87(supplement): 612–13.

Narrigan, Deborah, Jane Sprague Zones, Nancy Worcester, and Maxine Jo Grad. 1997. "Research to Improve Women's Health: An Agenda for Equity." In *Women's Health: Complexities and Differences,* edited by Sheryl Burt Ruzek, Virginia L. Olesen, and Adele E. Clarke, 551–579. Columbus: Ohio State University Press.

National Center for Chronic Disease. 2003. "Defining Overweight and Obesity." Center for Disease Control. September 23. Availble at: http://www.cdc.gov/nccdphp/dnpa/obesity/ defining.htm.

National Commission for the Protection of Human Subjects of Biomedical and Behavioral Research. 1979. *The Belmont Report: Ethical Principles and Guidelines for the Protections of Human Subjects of Research.* Washington D.C.: U.S. Department of Health, Education and Welfare.

National Institutes of Health Revitalization Act of 1993. Public Law 103-43 [S. 1]. 103rd Congress, 10 June.

Neugarten, Bernice L. 1974. "Age Groups in American Society and the Rise of the Young Old." *Annals of the American Academy of Political and Social Science* 415: 187–99.

Nobles, Melissa. 2000. *Shades of Citizenship: Race and the Census in Modern Politics*. Stanford, CA: Stanford University Press.

Nolan, Mary. 1994. *Visions of Modernity: American Business and the Modernization of Germany*. New York: Oxford University Press.

Official Journal of the European Union. 2003. C 52 E/132; written question E-2164/02, 6.3.2003. Daniel Hannan's question to the European Commission. January 18. Available at: http://www.ojec.com.

Olson, Gary M., Daniel E. Atkins, Robert Clauer, Terry Weymouth, Atul Prakash, Thomas A. Finholt, Farnam Jahanian, and Craig Rasmussen. 2001. "Technology to Support Distributed Team Science: The First Phase of the Upper Atmospheric Research Collaboratory (UARC)." In *Coordination Theory and Collaboration Technology*, edited by Gary M. Olson, Thomas W. Malone, and John B. Smith. Hillsdale, N.J.: Lawrence Erlbaum.

O'Neill, Robert V. 2001. "Is It Time to Bury the Ecosystem Concept? (With Full Military Honors, Of Course!)" *Ecology* 82(12): 3275–84.

Orloff, Ann Shola. 1988. "The Political Origins of America's Belated Welfare State." In *The Politics of Social Policy in the United States*, edited by Margaret Weir, Ann Shola Orloff, and Theda Skocpol, 37–80. Princeton: Princeton University Press.

Palme, Jacob. 1997. "Can Computers Decide What Is Right and Wrong?" Available at: http://dsv.su.se/jpalme/reports/right-wrong.html.

Palme, Sven. 1927. "The Insurance of Under-Average Lives." In *Transactions of the 8th International Congress of Actuaries, London 27th to 30th June, 1927*, vol. 2, compiled by S. H. Jarvis, 151–60. London: Layton.

Parker, Richard. 1995. *Mixed Signals: The Prospects for Global Television News*. New York: Twentieth Century Fund Press.

Pelling, Margaret. 1991. "Old Age, Poverty and Disability in Early Modern Norwich: Work, Remarriage and Other Expedients." In *Life, Death, and the Elderly: Historical Perspectives*, edited by Margaret Pelling and Richard M. Smith, 74–101. London: Routledge.

Pelling, Margaret, and Richard M. Smith. 1991. "Introduction." In *Life, Death, and the Elderly: Historical Perspectives*, edited by Margaret Pelling and Richard M. Smith, 1–38. London: Routledge.

Pick, Daniel. 1989. *Faces of Degeneration: A European Disorder, c. 1848–1918*. Cambridge, UK: Cambridge University Press.

Pittaway, Mark. 1999. "The Social Limits of State Control: Time, the Industrial Wage Relation, and Social Identity in Stalinist Hungary, 1948–1953." *Journal of Historical Sociology* 12(3): 271–301.

Poëls, E. 1927. "Les Risques Tarés." In *Transactions of the 8th International Congress of Actuaries, London 27th to 30th June, 1927*, vol. 3, compiled by S.H. Jarvis, 277–89. London: Layton.

Pollack, Andrew. 1998. "Drug Testers Turn to 'Virtual Patients' as Guinea Pigs." *New York Times*, 10 November, D1, D10.

———. 2002. "Drug to Treat Bowel Illness Is Approved by the F.D.A." *New York Times*, 25 July, A12.

Porter, Theodore M. 1986. *The Rise of Statistical Thinking*. Princeton: Princeton University Press.

——. 1995. *Trust in Numbers: The Pursuit of Objectivity in Science and Public Life*. Princeton: Princeton University Press.

——. 2000. "Life Insurance, Medical Testing, and the Management of Mortality." In *Biographies of Scientific Objects*, edited by Lorraine Daston, 226–47. Chicago: University of Chicago Press.

——. 2003. "Statistics and Statistical Methods." In *Cambridge History of Science*, vol. 7, The Modern Social Sciences, edited by Theodore M. Porter and Dorothy Ross, 238–50. Cambridge, UK: Cambridge University Press.

Prescott, Heather Munro. 2002. "Using the Student Body: College and University Students as Research Subjects in the United States During the Twentieth Century." *Journal of the History of Medicine* 57:3-38.

Proctor, Robert N. 1988. *Racial Hygiene: Medicine under the Nazis*. Cambridge, MA: Harvard University Press.

Rabinbach, Anson. 1990. *The Human Motor: Energy, Fatigue, and the Origins of Modernity*. Berkeley: University of California Press.

Raith, Tivadar. 1930. "A racionálizálásáról" [On Rationalization]. *Magyar Szemle* 10(11): 248–58.

Ratto, Matt. 2001. "Leveraging Software, Advocating Ideology: Linux, Free Software, and Open Source." Available at: http://www.arxiv.org/abs/cs.CY/0109103.

Raynes, Harold E. 1964. *A History of British Insurance*. London: Pitman.

Reichenbach, Béla. 1930. *Mezőgazdasági üzemtan* [Agricultural Science of Management]. Budapest: "Pátria" Irodalmi Vállalat és Nyomdai Részvénytársaság.

Republique Democratique du Congo. 1970. *Supplement au Codes Congolais*. Brussels: Maison Ferdinand Larcier, S.A.

Révai Nagy Lexikona. 1911. Vol. 1, s.v. "amerikázás." Budapest: Révai Testvérek Irodalmi Intézet R.T.

Révay, József. 1946. "A munka és szórakozás lélektana" [The Psychology of Work and Leisure]. *Lélektani Tanulmányok* 8: 86–107.

Rézler, Gyula. 1944. *A Magyar Ipari Munkatudományi Intézet* [The Hungarian Industrial Work Science Institute]. (A Magyar Ipari Munkatudományi Intézet 3. számú kiadványa.) Budapest: A Magyar Ipari Munkatudományi Intézet Kiadása.

Ritter, Gerhard A. 1998. *Social Welfare in Germany and Britain*. New York: Berg.

Robbins-Roth, Cynthia. 1998. "Tailor-Made Drugs: The Field of Pharmacogenomics Promises to Produce Drugs That Always Will Be Effective for an Individual." *Forbes*, 6 July, 172.

Rose, Nikolas. 1990. *Governing the Soul: The Shaping of the Private Self*. London: Routledge.

Rothman, David J. 1991. *Strangers at the Bedside*. New York: Basic Books.

Rothstein, Mark A., and Phyllis Griffin Epps. 2001. "Ethical and Legal Implications of Pharmacogenomics." *Nature Reviews: Genetics* (March): 228-231.

Rothstein, William G. 2003. *Public Health and the Risk Factor: A History of an Uneven Medical Revolution*. Rochester, NY: University of Rochester Press.

Rubinow, Isaac. M. 1913. *Social Insurance with Special Reference to American Conditions*. New York: Henry Holt and Company.

Rudolph, Karl. 1927. "Die Versicherung minderwertiger Leben in Deutschland." In *Transactions of the 8th International Congress of Actuaries, London 27th to 30th June, 1927*, vol. 4, compiled by S. H. Jarvis, 322–29. London: Layton.

Sanders, Heywood T. 1980. "Paying for the 'Bloody Shirt': The Politics of Civil War Pensions." In *Political Benefits: Empirical Studies of American Public Programs*, edited by Barry S. Rundquist, 137–59. Lexington, Mass.: Lexington Books.

Saul, Stephanie. 2005. "F.D.A. Approves a Heart Drug for African-Americans." *New York Times*, 24 June, C2.

Schauer, Thomas. 2003. "The Sustainable Information Society: Visions and Risks." Research Institute for Applied Knowledge Processing, project Strategic Action for a Sustainable Knowledge and Information Society (SASKIA), the European Commission IST programme, Universitätsverlag Ulm GmbH.

Schiebinger, Londa. 1993. *Nature's Body: Gender in the Making of Modern Science*. Boston: Beacon Press.

———. 2004. "Human Experimentation in the Eighteenth Century: Natural Boundaries and Valid Testing." In *The Moral Authority of Nature*, edited by Lorraine Daston and Feranando Vidal, 384–408. Chicago: University of Chicago Press.

Schiller, Pál. 1940. "Képességvizsgálat és munkateljesítmény" [Studies of Intellect and Work Performance]. *Lélektani Tanulmányok* 4: 159–75.

Schlosser, Eric. 2001. *Fast Food Nation*. Boston: Houghton Mifflin.

Schmucker, Douglas L. 1984. "Drug Disposition in the Elderly: A Review of the Critical Factors." *Journal of the American Geriatric Society* 32 (2):144-149.

Schmuhl, Hans-Walter 1987. *Rassenhygiene, Nationalsozialismus, Euthanasie: Von der Verhütung zur Vernichtung "lebensunwerten Lebens": 1890–1945*. (Kritische Studien zur Geschichtswissenschaft 75.) Göttingen: Vandenhoeck und Ruprecht.

Schrikker, Sándor. 1942. A sommások alkalmazásának és kereseti viszonyainak üzemi vonatkozásai [The Operative Aspects of the Employment and Income Relations of Migrant Workers]. PhD diss., M. Kir. József Nádor Műszaki és Gazdaságtudományi Egyetem.

Schumpeter, Joseph. 1934. *The Theory of Economic Development*. Cambridge, Mass.: Harvard University Press.

Schwartz, Robert S. 2001. "Racial Profiling in Medical Research" (Editorial). *New England Journal of Medicine* 344(18): 1392-1393.

Science Meets Reality: Recruitment and Retention of Women in Clinical Studies and the Critical Role of Relevance 2003. [online summary]. National Institutes of Health Office of Research on Women's Health [cited 25 April 2003]. Available from http://www4.od.nih.gov/orwh/smr.html.

Scott, Christopher, and Georges Sabagh. 1970. "The Historical Calendar as a Method of Estimating Age: The Experience of the Moroccan Multi-Purpose Sample Survey of 1961–63." *Demography* 24: 93–109.

Scott, James C. 1998. *Seeing Like a State: How Certain Schemes to Improve the Human Condition Have Failed*. New Haven: Yale University Press.

The Seventh Report of the Joint National Committee on Prevention, Detection, Evaluation, and Treatment of High Blood Pressure. 2003. Bethesda, MD: National Heart Lung and Blood Institute, National Institutes of Health, U.S. Department of Health and Human Services.

Shakespeare, William. 1600. *As You Like It. Accessed at "Quotes from As You Like It,"* compiled by Amanda Mabillard. *Shakespeare Online*. 2003. htt://www.shakespeare-online.com/quotes/ayliquotes.html.

Sharkey, Joe. 2002. "Airline Tries to Quell Seating Storm." *New York Times*, 25 June, C9.

Shryock, Henry S., and Jacob S. Siegel. 1973. *The Methods and Materials of Demography*. Washington, D.C.: U.S. Government Printing Office.

Skocpol, Theda. 1992. *Protecting Soldiers and Mothers*. Cambridge, Mass.: Harvard University Press.

Slaton, Amy, and Janet Abbate. 2001. "The Hidden Lives of Standards: Technical Prescriptions and the Transformation of Work in America." In *Technologies of Power*, edited by Michael Allen and Gabrielle Hecht, 95–143. Cambridge, Mass.: MIT Press.

Smocovitis, Vassiliki. 1996. *Unifying Biology: The Evolutionary Synthesis and Evolutionary Biology*. Princeton: Princeton University Press.

Social Security Administration. 1960. "Oldest Record Disagrees with Convincing, More Recent Evidence of Age SSR 60–17." *Cumulative Bulletin* 1960–1965: 77.

———. 1965. "Proof-of-Age—Effect of Court Decree on Finding of Secretary, SSR 65-34c Sections 202(c) and 205(g)." *Cumulative Bulletin* 1960–1965: 80.

———. 1981. "Titles II, XVI, and XVIII: Proof of Age Requirements for Holocaust Survivors SSR 81-16," *Cumulative Edition* 1981.

———. 2001. "Social Security Handbook Section 1705: How Do You Prove Your Age?" Available at: http://www.ssa.gov/OP_Home/handbook/handbook.17/handbook-1705.html. Accessed December 24, 2002.

Star, Susan Leigh. 1991. "Power, Technologies and the Phenomenology of Standards: On Being Allergic to Onions." In *A Sociology of Monsters?: Power, Technology and the Modern World*, edited by John Law, 26–56. London: Routledge.

———, ed. 1995. *Ecologies of Knowledge: Work and Politics in Science and Technology*. Albany: SUNY Press.

Star, Susan Leigh, and Geoffrey C. Bowker. 2002. "How to Infrastructure." In *Handbook of New Media: Social Shaping and Consequences of ICTs*, edited by Leah A. Lievrouw and Sonia Livingstone, 151–162. London: Sage Publications.

Star, Susan Leigh, and James Griesemer. 1989. "Institutional Ecology, 'Translations,' and Boundary Objects: Amateurs and Professionals in Berkeley's Museum of Vertebrate Zoology, 1907–1939." *Social Studies of Science* 19: 387–420.

Star, Susan Leigh, and Karen Ruhleder. 1996. "Steps toward an Ecology of Infrastructure: Design and Access for Large Information Spaces." *Informations Systems Research* 7(1): 111–34.

Stern, Alexandra Minna. 2002. "Making Better Babies: Public Health and Race Betterment in Indiana, 1920–1935." *American Journal of Public Health* 92(5): 742–52.

———. 2003. "From Mestizophilia to Biotypology: Racialization and Science in Mexico, 1920–1960." In *Race and Nation in Modern Latin America*, edited by Nancy P. Applebaum, Anne S. Macpherson, and Karin Alejandra Rosemblatt, 187–210. Chapel Hill: University of North Carolina Press.

Stévenin, H., 1968. "History of Aggravated Risks in Insurance." In *International Conference of Life Assurance Medicine in Tel Aviv 1967*, edited by Max Leffkowitz and H. Steinitz, 315–29. Basel.

Stigler, Stephen M. 1986. *The History of Statistics: The Measurement of Uncertainty before 1900*. Cambridge, Mass.: Harvard University Press.

St. John, Warren. 2003. "On the Final Journey, One Size Doesn't Fit All." *New York Times*, 28 September, A1.

Stoler, Ann. 1992. "Sexual Affronts and Racial Frontiers: European Identities and the Cultural Politics of Exclusion in Colonial Southeast Asia." *Comparative Studies in Society and History* 34(3): 514–51.

Strauss, Anselm. 1993. *Continual Permutations of Action*. New York: Aldine de Gruyter.

Sturm, Josef. 1935. "Methoden der Risikobewertung." In *International Congress of Life Assurance Medicine*, 29–81. London.

Summerton, Jane, ed. 1994. *Changing Large Technical Systems*. Boulder: Westview Press.

Szabó, Zoltán. 1937. *A tardi helyzet* [The Situation in Tard]. Budapest: Cserépfalvi Könyvkiadó.

Szakáll, Sándor. 1943. "Irányelvek a kaszás normális munkájának megállapitására" [Principles for Determining Normal Work for Mowing]. *Mezőgazdasági Munkatudomány* 1(July–August): 316–27.

Thane, Pat. 2000. *Old Age in English History: Past Experiences, Present Issues*. Oxford: Oxford University Press.

Thévenot, Laurent. 1984. "Rules and Implements: Investment in Forms." *Social Science Information* 23(1): 1-45.

Thompson, Stuart. 1990. "Metaphors the Chinese Age By." In *Anthropology and the Riddle of the Sphinx: Paradoxes of Change in the Life Course*, edited by Paul Spencer, 102–20. London: Routledge.

Thurzó, Jenő. 1934. "Psychologiai és bioszociologiai észrevételek az amerikai embertypusról és az 'amerikanizmus'-ról" [Psychological and biosociological observations on the American human type and "Americanism"]. Orvosok és Gyógyszerészek Lapja, klny.

Timmermans, Stefan. 1998. "Mutual Tuning of Multiple Trajectories." *Symbolic Interaction* 21(4): 425–40.

Timmermans, Stefan, and Marc Berg. 2003. *The Gold Standard: The Challenge of Evidence-Based Medicine and Standardization in Health Care*. Philadelphia: Temple University Press.

Tokarsi, Cathy. 2003. "For-Profit Research More Favorable in Drug Trials." *Medscape Medical News*. Available at: http://www.medscape.com/viewarticle/460329. Accessed November 5, 2003.

Treas, Judith. 1986. "The Historical Decline in Late-Life Labor Force Participation in the United States." In *Age, Health, and Employment*, edited James E. Birren, Pauline K. Robinson, and Judy E. Livingston, 158–75. Englewood Cliffs, N.J.: Prentice Hall.

Trigg, Randall, and Susanne Bødker. 1994. "From Implementation to Design: Tailoring and the Emergence of Systematization in CSCW." In *Proceedings of ACM 1994 Conference on Computer-Supported Cooperative Work*, edited by Richard Furata and Christine Neuwirth, 45–54. New York: ACM Press.

Troyansky, David G. 1989. *Old Age in the Old Regime*. Ithaca: Cornell University Press.

Turda, Marius. 2007. "The First Debates on Eugenics in Hungary, 1910–1918. In *Blood and Homeland: Eugenics and Racial Nationalism in Central and Southeast Europe 1900–1940*, edited by Marius Turda and Paul Weindling, 185–221. Budapest: Central European University Press.

Turda, Marius, and Paul Weindling, eds. 2007. *Blood and Homeland: Eugenics and Racial Nationalism in Central and Southeast Europe 1900–1940*. Budapest: Central European University Press.

Turnbull, David. 1993. "The Ad Hoc Collective Work of Building Gothic Cathedrals with Templates, String, and Geometry." *Science, Technology and Human Values* 18(3): 315–43.

Udry, J. Richard, and R. L. Cliquet. 1982. "A Cross-Cultural Examination of the Relationship between Ages at Menarche, Marriage, and First Birth." *Demography* 19: 53–63.

Ujlaki, Nagy Árpád. 1943. "A mezőgazdasági munka mechanikai, termodinamikai és biológiai értéke" [The Mechanical, Thermodynamic and Biological Value of Agricultural Work]. *Mezőgazdasági Munkatudomány* 4 (July–August): 299–315.

U.S. Department of Agriculture Center for Nutrition Policy and Promotion. 2000. "Serving Sizes in the Food Guide Pyramid and on the Nutrition Facts Label: What's Different and Why?" *Nutrition Insights* 22, 23 September. Avaliable at: http://www.usda.gov/cnpp.

Verran, Helen. 2001. *Science and an African Logic.* Chicago: University of Chicago Press.

Wade, Nicholas. 2002. "Scientist Reveals Genome Secret: It's Him." *New York Times,* 27 April, A1.

Wagner, Adolph. 1881. *Der Staat und das Versicherungswesen.* Tübingen: Laupp.

Wagner, Peter. 2003. "The Uses of the Social Sciences." In *Cambridge History of Science,* vol. 7, *The Modern Social Sciences,* edited by Theodore M. Porter and Dorothy Ross, 537–52. Cambridge, UK: Cambridge University Press.

Wald, Matthew L. 2000. "Revised Rules for Air Bags Will Reduce Their Power." *New York Times,* 5 May, A14.

———. 2003. "Weight Estimates on Air Passengers Will Be Increased." *New York Times,* 13 May, A1.

Waldby, Catherine. 2000. *The Visible Human Project: Informatic Bodies and Posthuman Medicine.* London: Routledge.

Wallace, Peggy. 1997. "Following the Threads of an Innovation: The History of Standardized Patients in Medical Education." *Caduceus* 13(2): 5–28.

Wallraff, Barbara. 2000. "What Global Language?" *Atlantic Monthly* (November): 52–66.

Warner, John Harley. 1986. *The Therapeutic Perspective: Medical Practice, Knowledge, and Identity in America, 1820–1885.* Cambridge, Mass.: Harvard University Press.

Weick, Karl E. 1979. *The Social Psychology of Organizing.* 2nd ed. Reading, Mass.: Addison-Wesley.

Weingart, Peter, Jürgen Kroll, and Kurt Bayertz. 1988. *Rasse, Blut und Gene: Geschichte der Eugenik und Rassenhygiene in Deutschland.* Frankfurt: Suhrkamp.

Weis, István. 1930. *A mai magyar társadalom* [Modern Hungarian Society]. Budapest: Magyar Szemle Társaság.

Weisman, Carol S. 1998. *Women's Health Care: Activist Traditions and Institutional Change.* Baltimore: John Hopkins University Press.

———. 2000. "Breast Cancer Policymaking." In *Breast Cancer: Society Shapes an Epidemic,* edited by Anne S. Kasper and Susan J. Ferguson, 213–43. New York: St. Martin's Press.

Weiss, Rick. 2000. "The Promise of Precision Prescriptions; 'Pharmacogenomics' Also Raises Issues of Race, Privacy." *Washington Post,* 24 June, A1.

Weizenbaum, J. 1976. *Computer Power and Human Reasoning: From Judgment to Calculation.* W. H. Freeman and Company.

White, Leonard D. 1958. *The Republican Era: 1869–1901.* New York: Macmillan.

Wimsatt, William C. 1998. "Simple Systems and Phylogenetic Diversity." *Philosophy of Science* 65(2): 267–75.

Winterthur. 2000. *Kontakt: Ratgeber Datenschutz.* Winterthur: Winterthur-Versicherungen.

Worldlingo. 2000. "The WorldLingo Quarterly Email Survey—November 2000." Available at: http://www.worldlingo.com/resources/nov_email_survey.html.

———. 2001a. "The WorldLingo Quarterly Email Survey—April 2001." Available at: http://www.worldlingo.com/resources/apr_email_survey.html.

———. 2001b. "The WorldLingo Quarterly Email Survey—August 2001." Available at: http://www.worldlingo.com/resources/aug_email_survey.html.

Wright, Lawrence. 1994. "One Drop of Blood." *New Yorker,* 25 July, 46–55.

Yaffe, Sumner J., ed. 1980. *Pediatric Pharmacology: Therapeutic Principles in Practice.* New York: Grune & Stratton.

———. 1983. "Problems of Drug Testing in Children in the United States." *Pediatric Pharmacology* 3: 339–348.

Yates, JoAnne. 1989. *Control through Communication: The Rise of System in American Management.* Baltimore: Johns Hopkins University Press.

Young, Allyn A. 1904. *A Discussion of Age Statistics*, vol. 13. Washington, D.C.: Government Printing Office.

Young, Lisa R., and Marion Nestle. 2002. "The Contribution of Expanding Portion Sizes to the U.S. Obesity Epidemic." *American Journal of Public Health* 92: 246–49.

Zernike, Kate. 2004. "Sizing up America: Signs of Expansion from Head to Toe." *New York Times*, 1 March, A1.

CONTRIBUTORS

GEOFFREY BOWKER is the Regis and Dianne McKenna Professor and director of the Center for Science, Technology and Society at Santa Clara University.

ELIZABETH CULLEN DUNN is an associate professor in the Department of Geography and the International Affairs Program at the University of Colorado at Boulder.

STEVEN EPSTEIN is a professor of sociology and science studies, and the director of the Science Studies Program at the University of California, San Diego.

MARTHA LAMPLAND is an associate professor of sociology and science studies, and the director of the Critical Gender Studies Program at the University of California, San Diego.

MARTIN LENGWILER is a research associate in science policy studies at the University of Zurich.

FLORENCE MILLERAND is an assistant professor in the Department of Social and Public Communication at the University of Quebec at Montreal.

JACOB PALME is a professor of computer and systems sciences at Stockholm University.

DANIEL PARGMAN is an assistant professor in media technology at the School of Computer Science and Communication, Royal Institute of Technology (KTH), Sweden.

SUSAN LEIGH STAR is a research professor at the Center for Science, Technology and Society at Santa Clara University.

JUDITH TREAS is a professor of sociology at the University of California, Irvine.

INDEX

Page numbers in italics refer to figures, illustrations, and tables.

Becker, Howard S., 24
BenchMark Clothiers, 61
"Bias in Computer Systems" (Friedman
    and Nissenbaum), 193–95
Biddenden Maids, 29–30
BiDil (heart failure drug), 51
Biederman, Marcia, 58–62
Bijani, Laleh and Ladan, 28, 29
Billings, Paul G., 147
biomedical research, 41–42
    uncertainty of, 52–53
Birth Registration Act (1910), 76
Blakely, Robert L., 47–48
Blaschke, Ernst, 101
Bloss, Barry, 89
Bowker, Geoffrey, 6, 7, 16, 17, 22, 95,
    181–82, 189
British Clerical, Medical and General Life
    Assurance Company, 103–4
Brooks Brothers (clothing retailer), 60
Brown, Nik, 42
Browner, Carol, 147
Bucheim, Eduard, 100, 101, 108–9
Budd, John, 75
"A Bulge in Misses 8" (Biederman), 58–62
Bunker, Eng and Chang, 29, 30
Bureau of Land Management, U.S., 143–44
Busch, Lawrence, 95, 96
Bush, George W., 143–44, 148, 175, 176

California
    Three Strikes Law, 115, 117
California, drinking water standards, 146
Callon, Michel, 14
Carlin, George, 65
Cartwright, Lisa, 48
Cavaney, Red, 148
Census Bureau, U.S., 76, 77–79, 80–81, 83
Center for Responsive Politics, 148
Centers for Disease Control and
    Prevention, 89
character set standards, 177–95
    bias in, 193–95
    development of, 189–91
    implementation of, 191–92
    ISO 646, 182–83
    ISO 2022, 183
    ISO 8859, 183
    ISO 10646, 184
    language and, 179–80

non-ASCII characters, challenge of,
    184–86, *187–88*
    Unicode, 183–84
character set standards organizations,
    189, 190
Charleston Daily Mail, 89–90
China, chronological age and, 68–69
Chocolate Directive (2000, EP), 204–5
"Chocolate Directive Now Agreed"
    (UK Ministry of Agriculture Fisheries
    and Food), 204–5
chocolate standards, 204–5
chronological age, 65–87
    administrative breakdowns, 81–82,
        83–85
    age exactitude and, 75–76, 80–81
    Census Bureau and, 76, *77–79,* 80–81
    Civil War pensions and, 73–74
    metric for standards, 65–66
    old age and, 70–72
    pension systems and, 74–75
    proof of age documentation of, 82–83
    public assistance and, 71
    slow embrace of, 68–70
    as stratification mechanism, 86–87
    See also age
Civil War pensions
    chronological age and, 73–74
    eligibility criteria, 72–73
Clapp, Philip, 144
Clinton, Bill, 143, 144, 147, 148, 203
Clinton, Hillary Rodham, 174
Codex Alimentarius, 119–20
"Coffins Expand with Occupants" *(Charleston
    Daily Mail),* 89–90
Cohen, Jay S., 37, 49n9
Cointra (Coopération Internationale pour les
    Assurances sur la Vie des Risques
    Aggravés), 110
Cole, Thomas R., 70
Collins, Patricia Hill, 141
Communist Party (Hungary), 136, 137
    folk education campaigns, 139–40
Compagnie Belge d'Assurances Générales sur
    la Vie, 108
Comte, Auguste, 96
conjoined twins, 28–31
consumer goods standards, 36
conventions, comparison to standards,
    10n4, 24